3D CAD with Autodesk 123D

Designing for 3D Printing, Laser Cutting, and Personal Fabrication

Jesse Harrington Au and Emily Gertz

3D CAD with Autodesk 123D

by Jesse Harrington Au and Emily Gertz

Printed in the United States of America.

Published by Maker Media, Inc., 1160 Battery Street East, Suite 125, San Francisco, CA 94111.

Maker Media books may be purchased for educational, business, or sales promotional use. Online editions are also available for most titles (*http://safaribooksonline.com*). For more information, contact O'Reilly Media's institutional sales department: 800-998-9938 or *corporate@oreilly.com*.

Editor: Patrick Di Justo
Production Editor: Nicholas Adams
Copyeditor: Jasmine Kwityn
Proofreader: Charles Roumeliotis
Indexer: Judith McConville
Interior Designer: David Futato
Cover Designer: Brian Jepson
Illustrator: Rebecca Demarest

January 2016: First Edition

Revision History for the First Edition

2015-12-18: First Release

See *http://oreilly.com/catalog/errata.csp?isbn=9781449343019* for release details.

978-1-449-34301-9

[LSI]

Table of Contents

Introduction to the 3D Dimension

We assume you are reading this book because you've arrived at a stage in your creative life where you'd like to do more with your computer. You would love to combine its power with advances in computer-aided design (CAD) and fabrication to make something awesome in three dimensions.

Many makers who are accustomed to creating by hand view CAD software suspiciously. They may worry that digital design will lack soul, or be perceived as cheating. Neither is true. A good CAD program can be just that: an aid in realizing your vision for your project.

Just like any other tool, CAD is best viewed as a step on the way to your ultimate goal. You would not build a bench with a hammer as your only tool. So why would you try to design an object using only CAD? In fact, the best way to use a virtual design space involves always keeping an eye on the final step of physical fabrication. If you designed a table to be made from a single sheet of plywood eight-and-a-half-feet long, you might eventually drive yourself insane—because plywood sheets typically come in eight-foot lengths.

The same goes for transporting an object. We couldn't tell you how many times we've seen someone make an absolutely gorgeous piece of furniture, only to be crushed by the fact that it can't fit through a doorway.

Another important factor in design is sustainability. Making small changes to a design—for example, devising it in such a way that its different parts can all be cut from a single piece of material, or so that it doesn't require expensive fasteners or injection molding techniques—can save you a lot of money as well as the world a lot of wasted material that ends up in a landfill.

The idea for this book came from Jesse's daily work with people in a variety of maker spaces, educational facilities both traditional and non-traditional. The makers he met were eager to learn how to design for three dimensions using a computer, but had no idea where to start.

In this book, we'll talk about "putting a hole" in an object, rather than the technical term "perform a Boolean subtraction." Don't worry, though: you will still come away knowing a bit more of the professional CAD and 3D design terminology, the better to impress the people you tell about your rad project—and even more important, to communicate with fellow designers and fabricators such as yourself.

I (Jesse) began building things at a very young age. I recall my father sitting down with my brother and me, graph paper and a few pencils in hand, and asking us to help design a solar-heated water pump for our pool. Now, keep in mind that I grew up in upstate New York, where it feels as if there is no sun for 10 months out of the year. During the first of the other two months, swimming in that pool was like joining the Polar Bear Club.

I remember my father relating to us the importance of planning in advance for a construction project, especially taking measurements. We would sit for long hours every night, figuring out how much PVC pipe it would take to circulate the water through the heater and warm it efficiently before returning back to the pool. Two factors kept my brother and me attentive during these evenings: the prospect of warmer pool water, which would mean more pool parties; and the highest prize of all, being allowed to use a power saw on the roof of the house.

In the end, the project came together perfectly, resulting in a warm pool and great summers for our remaining years in that house. But what I took away long term was a love for the process, the endless puzzle solving that goes into good design. Will I need a jig to cut certain pieces? How many ways can I smarten up my design so that it will result in an object that is truly all it can be? How should I tweak it to get a final creation that I can ship easily?

These are the sorts of little details that I absolutely love to perfect while toiling away in the world of CAD: the little considerations that I've learned to inject into a design, after a lifetime of tinkering with poorly made bookshelves or other objects that were made to last a few months at best.

So when you get frustrated with the computer, or you forget to save your file, remember: this is the heart of design. Instead of coping with some other designer's bad decisions, now you get to make choices that will result in your object being the most memorable object it can be. When you create something—whether it be a 3D print or a handmade chair—you are putting a part of yourself into it. So make sure it's something that you can take pride in. Take a little extra time to worry about the radius of the curves or the correct hole sizes. Because if you don't, it just may haunt you for the rest of your life.

Things to Know Before You Get Started

In our experience, many new 3D designers and makers are hard-pressed to find projects that they would like to create. Great ideas come from many places, but if you are struggling to find inspiration, try looking for the gaps or hacks that people come up with to "fix" objects they use in daily life. Every time we see someone patch something together with duct tape, we can't help but think to ourselves that their situation could have been made easier by better design.

Also, keep an eye out for needs that would be well solved with an object made out of plastic. For example, Jesse remembers helping a college student who was working on generating power from a kite with an attached wind turbine. An easy solution would have been to use electrical tape to attach the turbine. But then, if the project failed, how would he know if the problem was faulty design or faulty tape? A better solution would be a 3D-printed brace or harness for the turbine.

These kinds of scenarios, which come up often for inventors, are when you'll be especially glad to minimize the variables by way of sturdily made, 3D-printed parts.

Which Program Is Right for Me?

There are two major categories for modeling programs in the 123D universe, surface and solid. *Surface models* are distinct in that they have no thickness, so what you see on the surface is what you get with the model. They are created with a triangulated mesh that makes up the surface.

Solid models have thickness, inserted by giving the program commands for measurement. A benefit of solid modeling is that you can go back and change those measurements. For instance, perhaps you created a piece of furniture and the joints you created were based on a 0.75-inch-thick piece of wood. But upon receiving your material, you find out that the wood is actually 0.77 inches thick. A solid modeling program allows you to go back in and change your measurement based on the desired amount.

Although the two types of programs, surface and solid, seem to get more alike with every release, there are still distinct differences between them. Some are based on how you think as a maker. Let's look at the reasons why you might want to choose one type of program over another.

If you want to create something mechanical or structural (or based on measurements), you should consider *123D Design*, or a similar program.

Solid modeling is what we think of when we think of engineers sitting in a dark room, creating complicated robotics or parts that need to be machined.

Solid models are almost always based on measurements (see "About Parametric Modeling"). They are also for things such as flat pack furniture design, fittings, tools, motors, phone cases, robots, usable plastic parts, and many other things that require precise measurements, or need to fit together.

With a bit of finesse, you can solid model just about anything. Some solid modeling programs, like Autodesk Inventor, SolidWorks, or PTC Credo (formerly Pro/Engineer), actually have sections that create surface models that can be implemented into your solid model. These are great for parts that need to be a bit more organically shaped, such as grips for handles or casings for plastic parts.

123D Design is capable of creating solid models that can also be very organic-looking in nature, and easily manipulated without having to create a different type of model. For example, see Figure P-1.

Figure P-1 *A 3D design of a helicopter*

If you want to create something that is based on a cool character, an organic shape, or something found in nature, you should consider *123D Catch*, *123D Meshmixer*, or *123D Sculpt+*.

Surface models (see "About Surface Modeling"), also called point cloud models, tend to be more organic in shape and are primarily used for everything from character design for toys, film, and animation to car bodies. For example, see Figure P-2.

Figure P-2 *A monster created using all the tools in the 123D suite*

About Parametric Modeling

Autodesk 123Design, Tinkercad, Autodesk Inventor, AutoCAD, SolidWorks, and PTC Credo are considered "parametric design" software. The term *parametric* refers to the use of design parameters, such as measurements, to construct and control the 3D model. This means you will first create a sketch that has measurements attached to it. Those measurements will be used to construct your solid model using different features such as extrude, revolve, or loft.

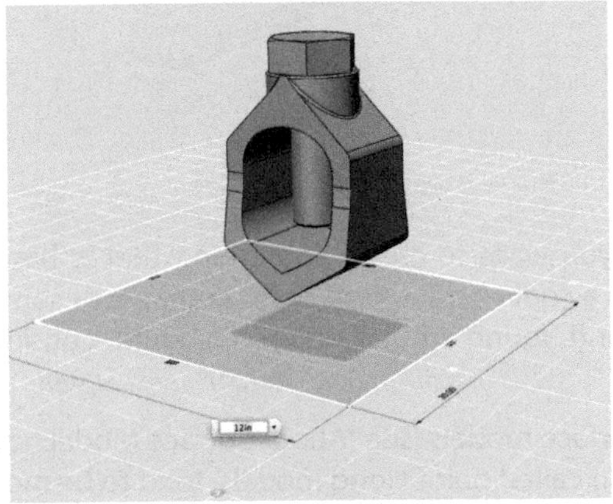

Figure P-3 *A mechanical part, ready to be 3D printed*

This being said, 123D is also capable of "tinkering": using loosely based measurements while fleshing out the look and feel of your design. The power of this is that it allows you to tweak your model during the design process based on actual measurements.

Figure P-4 *A simple birdhouse made in 123D*

Figure P-5 *Industrial containers, in designer colors*

About Surface Modeling

Surface models are normally used hand in hand with solid models. These programs do have some parametrics built in, but they also have the ability to create very organic-looking parts.

When designers use a program like Autodesk Alias or Rhino, or the surface modeling side of Autodesk Inventor, Fusion 360, or SolidWorks, they are literally only creating the very surface of an object. Put another way, they are creating geometry that has zero thickness.

This allows them to create shapes that would be difficult to create using solids. A good example of this is a mouse (the clickity-click kind, not the squeaky-squeak kind). The compound curves found on a mouse are sometimes difficult to model using the solid modeling tool chains. So, what many software companies often do is create a section of their modeling software that just creates surfaces. Then those surfaces can be run through another tool that gives them thickness, and thus makes them solid.

It's confusing, I know. The upshot is the combination allows for greater flexibility when it comes to creating specific curves. If you were designing a car, for instance, and you wanted a very specific curvature on its surface to give a visual impression of speed, that's where surface modeling would come into play.

There are two types of surface model: nurb based and point cloud based. The term *nurb*, beyond being really fun to say, stands for *non-uniform rational B-spline*. What does that mean, and what the heck is a spline?

Think of a spline like a guitar string attached to a nail that's hammered into a wooden board. Stretch the string taut, parallel to the surface of the board, and begin to turn it in a circle. Each time you pound in another nail and pull the guitar string against it, the string curves around that point.

Splines exist in both surface and solid modeling as a way to make delicate curves out of what is, essentially, a series of lines between points.

There's also a type of surface modeling program that uses something called a T-spline. Primarily this software is used for plastic part or automotive design. Programs that typically fall into this category include Rhino, SketchUp, and parts of Autodesk Fusion 360.

One issue with these programs is that they don't always create solid models—meaning that the files they produce are not always watertight for printing. *Watertight* is when all of the little triangles that make up the surface of your model (called tessellations) are lined up so there are no holes. This is important, because if there is a hole, your model might not print at all or it might print a solid brick instead of the desired shape. So if 3D printing or CNC is your final fabrication method, Rhino and SketchUp can be problematic for this reason, while Autodesk Fusion 360 can create a watertight solid.

That said, I have seen some amazing things built from each program. However, to get a watertight object, a new user may have to take designs created with Rhino or SketchUp into an additional digital workspace, such as Meshmixer, 123D 3D print utility, or Magix, to seal up the mesh before taking the design to fabrication.

Navigating a 3D Workspace

Most of us are accustomed to reading and drawing on the computer in two dimensions, or on a flat surface such as a piece of paper. So learning to get around in a 3D workspace can be a bit tricky.

Just remember—it's as simple as x, y, and z:

- The *x*-axis runs from left to right as you face forward on a machine or look at the front view of an object on a computer screen.
- The *y*-axis runs from front to back. Think of it as running from the space bar to the Y key on the keyboard.
- The *z*-axis runs up and down, between the "top" and the "bottom" of the design.

When starting a drawing, you'll begin it along two axes, referred to as the *plane*. These planes are infinitely long as far as the workspace is concerned, no matter how small they appear in your CAD program—something important to keep in mind.

Think of the *xy*-plane as the build platform of the 3D printer you intend to use. Just as you'd keep the size of a standard sheet of plywood in mind when designing a plywood tabletop, you'll want to keep the size of the build platform firmly in mind as you create your 3D-printable object.

To visualize the *xz*-plane, imagine that there's a wall directly between your eyes that stretches far out in front of you, from above your head down to your toes.

Think of the *yz*-plane as a wall that's standing in front of you, or, as running parallel to the computer screen, from one side to the other.

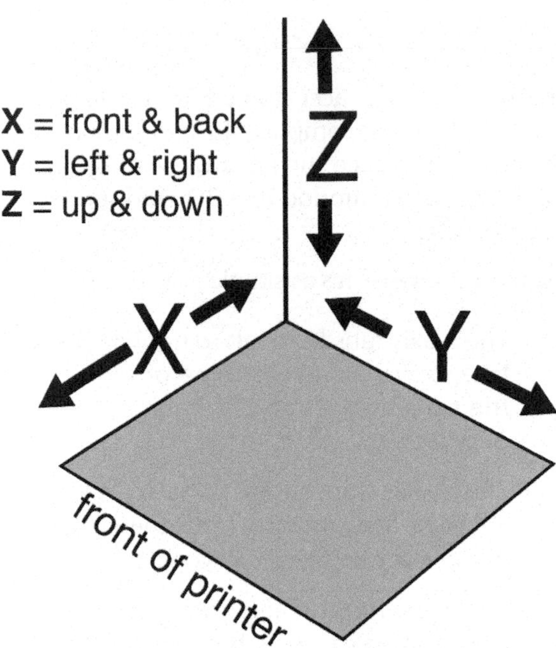

Figure P-6 *The mutually perpendicular x, y, and z planes*

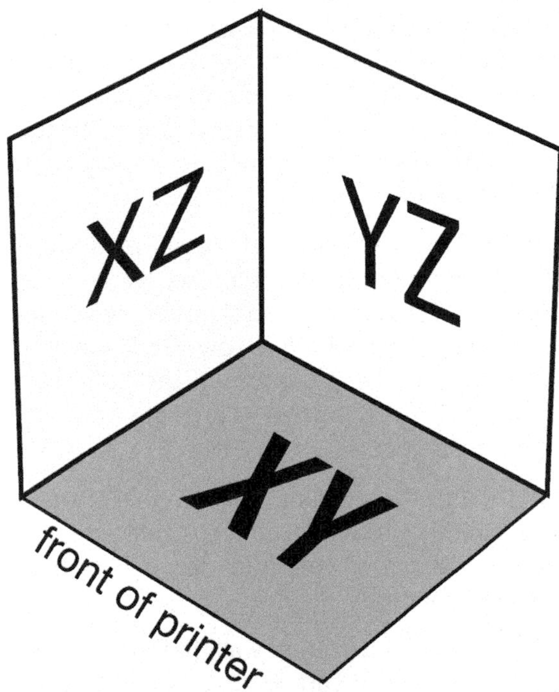

Figure P-7 *The mutually perpendicular x, y, and z axes*

The Three Ps of CAD

In closing, I'd like to mention the *Three Ps of CAD*: pain, psychological torture, and punishment. Oops...I meant to say patience, practice, and persistence.

Patience, in this case, refers to the ability to train yourself as a user and designer, as much as it does the ability to train yourself on how to use the program. Think about training a new pet: often we're actually training ourselves on how to communicate with the animal. But at least dogs haven't changed fundamentally in our lifetimes, whereas computers have become so powerful and ubiquitous, so quickly, that our expectations for how they "should" function are sometimes unrealistic. The same can be said for CAD, so be patient with yourself and with the design programs you explore.

Practice with CAD is essential. Working in 3D and creating with a mouse is very different than the other point-and-click that most of us do on a computer or the Internet. You will be training both your mental processes and your muscle memory. Also, you'll most likely crash a program the first few times you use it. That is in part due to faulty engineering of the program, and in part because you may be trying to do something too difficult for the program to process. The larger point here is this: don't give up. It will take time to get to a point where you are working smoothly, without disruption.

That brings us to the last P: *persistence*. When a design seems to go wrong, or a program gets difficult, you may be tempted to find out just how far a new laptop will go when thrown overhand out a window. But trust us: we guarantee that while it will be frustrating and difficult, you can do this.

Good luck, have fun, and blow something up. Safety third!

Conventions Used in This Book

This element signifies a general note, tip, or suggestion.

This element indicates a warning or caution.

Safari® Books Online

Safari Books Online is an on-demand digital library that delivers expert content in both book and video form from the world's leading authors in technology and business.

Technology professionals, software developers, web designers, and business and creative professionals use Safari Books Online as their primary resource for research, problem solving, learning, and certification training.

Safari Books Online offers a range of plans and pricing for enterprise, government, education, and individuals.

Members have access to thousands of books, training videos, and prepublication manuscripts in one fully searchable database from publishers like O'Reilly Media, Prentice Hall Professional, Addison-Wesley Professional, Microsoft Press, Sams, Que, Peachpit Press, Focal Press, Cisco Press, John Wiley & Sons, Syngress, Morgan Kaufmann, IBM Redbooks, Packt, Adobe Press, FT Press, Apress, Manning, New Riders, McGraw-Hill, Jones & Bartlett, Course Technology, and hundreds more. For more information about Safari Books Online, please visit us online.

How to Contact Us

Please address comments and questions concerning this book to the publisher:

> Make:
> 1160 Battery Street East, Suite 125
> San Francisco, CA 94111
> 877-306-6253 (in the United States or Canada)
> 707-639-1355 (international or local)

Make: unites, inspires, informs, and entertains a growing community of resourceful people who undertake amazing projects in their backyards, basements, and garages. Make: celebrates your right to tweak, hack, and bend any technology to your will. The Make: audience continues to be a growing culture and community that believes in bettering ourselves, our environment, our educational system—our entire world. This is much more than an audience, it's a worldwide movement that Make: is leading—we call it the Maker Movement.

For more information about Make:, visit us online:

- Make: magazine: *http://makezine.com/magazine/*
- Maker Faire: *http://makerfaire.com*
- Makezine.com: *http://makezine.com*
- Maker Shed: *http://makershed.com/*

We have a web page for this book, where we list errata, examples, and any additional information. You can access this page at *http://bit.ly/3DCADwithAutodesk123D*.

To comment or ask technical questions about this book, send email to *bookquestions@oreilly.com*.

How to Navigate Any CAD Program...Ever

1

123D Design is easier to use than most high-end CAD programs. But it has a similar layout to more complex programs, one that is found in most CAD programs that you will come across.

The basic elements of this layout are:

- The view cube
- The ribbon or control panel
- Mouse controls
- The right side menu
- A design tree or history of changes to the design

View Cube

The view cube (see Figure 1-1) is identical in the majority of Autodesk programs, as well as other CAD programs. Located in the upper-right corner, this tool makes it easy to get back to a standard view of your design. The standard views are very similar to the *xyz*-space described in "Navigating a 3D Workspace".

Figure 1-1 *The view cube*

What many new users don't realize is that the view cube, when placed on a specific view of the model, such as one of the views shown in Figure 1-2, is a great way to ensure that your object is lined up correctly. For example, when you are trying to use a move command or creating a 3D feature, the view cube can help you place items by first aligning them in a side view, and then aligning them in a top view, thus securing them in their *xyz*-space.

Remember that you are creating in 3D space, and moving something a small distance in one direction can result in a large distance in another direction, if you're not careful.

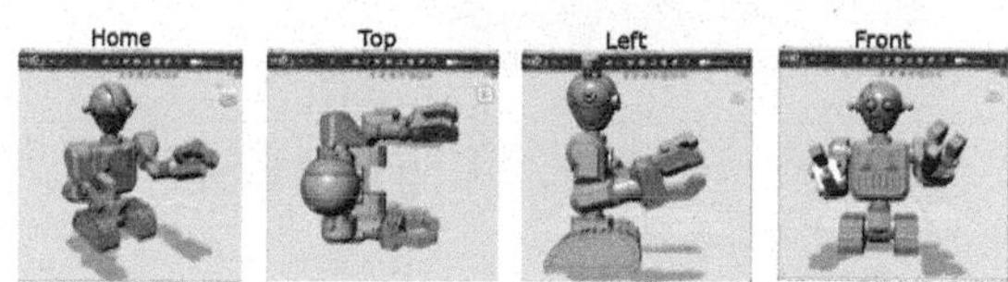

Figure 1-2 *Various views you can set from the view cube*

Ribbon

Located on the top middle of the screen, the ribbon (Figure 1-3) or control panel is the space where most of your tools are held. Just like the view cube, the ribbon is becoming standard in a variety of CAD products. We'll cover what these tools do later in the book.

Figure 1-3 *The ribbon, or control panel*

Main Menu

The main menu (Figure 1-4), as in almost any other app or program that you have ever used on a computer, is located in the upper-left corner of the workspace window. It is usually a drop-down menu, accessed by clicking on an icon that represents the product.

The main menu usually contains common commands such as Open, Save, Save a Copy, and Export. Special options for the particular program will generally be hidden in here as well.

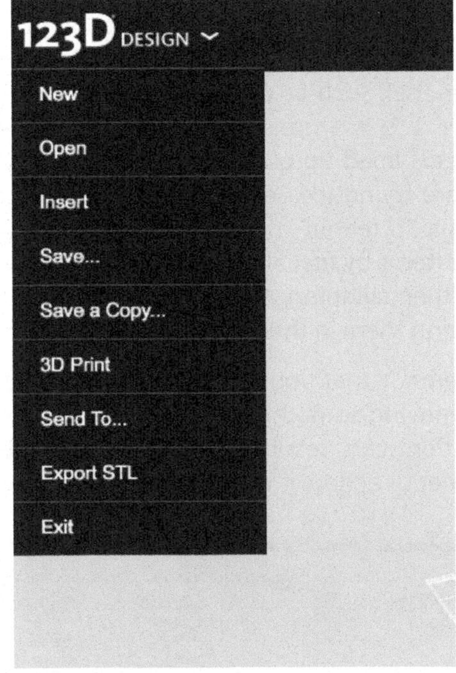

Figure 1-4 *The main menu*

Help Menu

The help menu (Figure 1-5) is located in the upper-right corner of your screen and represented by the question mark symbol. Click on this to see a submenu of helpful features, such as quick start tips and video tutorials.

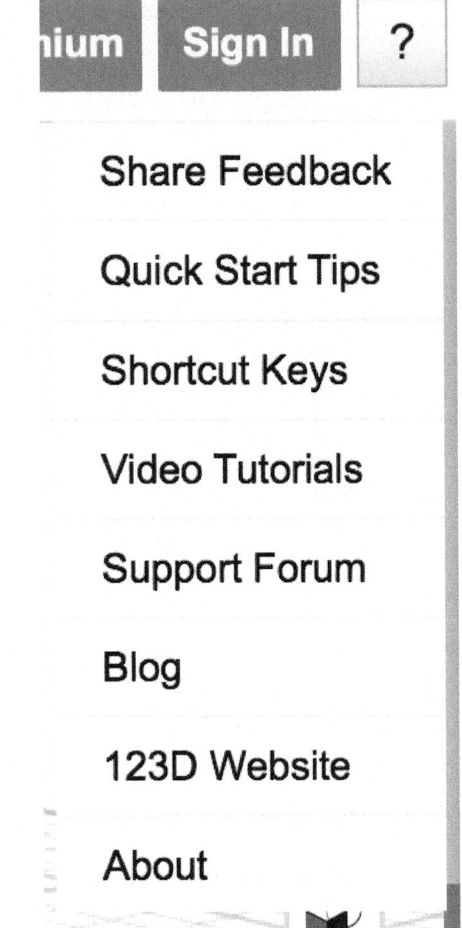

Figure 1-5 *The help menu*

Sign In

123D is a cloud-based family of programs that allow you to share models between the different apps. As shown in Figure 1-6, use the Sign In feature located near the help menu to create and then access your own account, which you can then use to save your own creations to the cloud, as well as explore projects shared by other users.

Remember: 123D is growing every day. Like any other relationship, you need to be involved to grow with it.

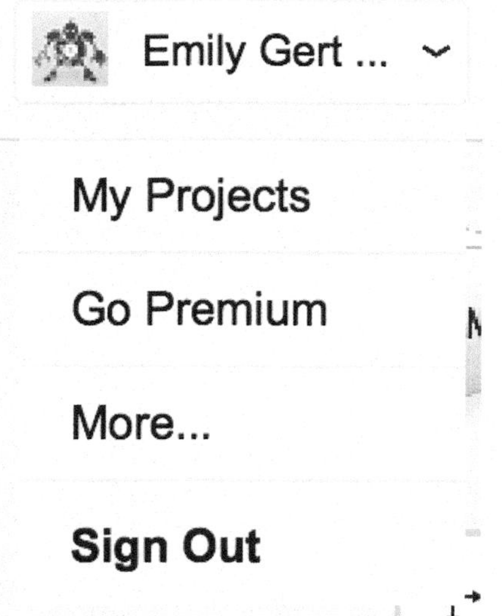

Figure 1-6 *Signed into 123D*

Mouse Controls/View Controls

It's a good idea to use a three-button mouse (i.e., a mouse with a right button, a left button, and a clickable scroll wheel) when working with any CAD program. These buttons are essential in unlocking all the menu systems. In addition, a three-button mouse will significantly lower your frustration levels.

You may be tempted to design using the computer's trackpad. However, for your sake and the sake of your loved ones, don't do it. Why? Most CAD programs, 123D included, use all three buttons for different functions. For example, a right-click will sometimes open up a menu that would be difficult to access in other ways.

The other reason is that if you attempt to use the trackpad, your fingers will crumple up like dried chicken feet, and you will have a hard time fabricating the things that you design.

In 123D Design for the desktop, you simply use your three-button mouse to navigate, select, and drag. For other CAD programs, you may have to use an additional function key to assist you. In Autodesk Inventor, for instance, you need to use the Shift key plus the Command key to unlock menus that can speed up your design work.

Select

Use your left mouse button to select an object. The first click will select the object, and the second click will select either a face or edge of the object.

Select Multiple

Depending on what tool you are using, you can either hold the Shift key down while you are making a selection to select multiple objects or simply select multiple objects.

Orbit

Use your right mouse button to rotate an object.

Orbiting does not actually rotate the object—instead, it simply rotates your view of the object. To rotate your object, select your object and then use the move command.

The Right Side Menu

The right side menu (Figure 1-7) that appears in most CAD workspaces will almost always contain the following controls: pan, orbit, zoom, and fill the screen/fit to view. Let's take a look at each of these.

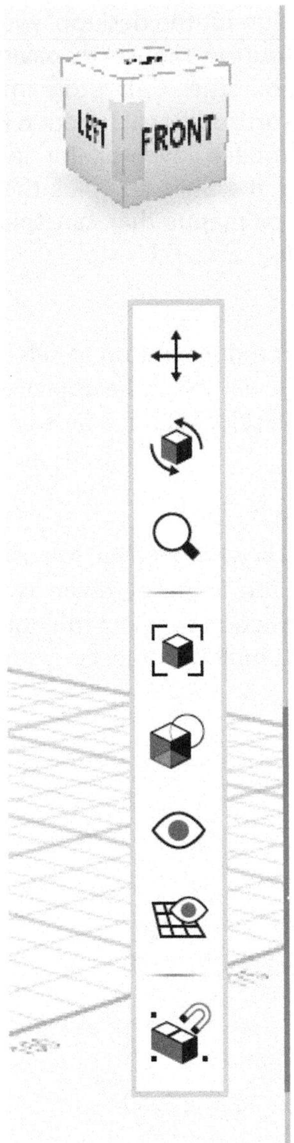

Figure 1-7 *The right side menu in 123D Design*

Pan

To pan, hold the middle mouse button down and move it left or right, or up or down.

Orbit

Move the on-screen cursor so that it is on or near the model. Then, click and hold the middle mouse button. Move the mouse in any direction you desire to rotate your view of the model.

Zoom

Roll the middle mouse to zoom into and out from your model.

Fill the Screen/Fit to View

Click and hold the middle mouse button to have your object fill the screen—great for working on small details.

> **Space Mouse!!!**
>
> A space mouse is a special kind of mouse that hardcore geeky types use to make things go a bit faster. If you find yourself falling deeply in love with CAD and think you might spend weeks of your life in the pure joy of virtual space creation, we highly recommend picking up a space mouse. Everyone we know who has used one says they would never again touch a CAD program without it. The space mouse looks like a hockey puck. It's easy to hold, and when you move it around, your 3D model moves in corresponding ways. It is a fantastic way to have your orbiting and zooming commands in your left hand while having your selection tools in your right.

Design Tree

Some higher-end CAD programs have what is called a design tree, usually located on the left side of the screen. This tree is a history of sorts of the design, calling out the features or bodies of what you've created.

The power of a design tree is that it allows the designer to go back and change very specific features of the design. In 123D Design, simply select the face or edge of what you would like to change and alter it via the gear menu that pops up after selection.

There are also undo and redo buttons located on the ribbon, which make it easy to take a quick step back from a change.

123D Sculpt+

2

123D Sculpt+ is our version of an evil scientist's laboratory. It's a beginner-friendly, free character-modeling app for use on iPad or an Android tablet. If you don't have an iPad or Android tablet, try using the sculpting features in 123D Meshmixer, covered later in this book, to do similar work.

In Sculpt+ you can create a skeleton, cover it with digital clay, shape and add texture to details like faces and limbs, and then export your creation for full-color 3D printing. You can do further design and detailing in pro modeling programs like Mudbox, Maya, 3ds Max, or ZBrush, and finally render it into live environments for sharing in the 123D Sculpt+ community.

To unlock all features in Sculpt+, create an account and sign in every time you use it.

Get Started with 123D Sculpt+

Beg, borrow, or buy an iPad or an Android tablet, and download the 123D Sculpt+ app.

When you first open Sculpt+, you'll see a welcome screen. Swipe to the left to see introductions to the five basic steps of the program:

- Create a skeleton/armature
- Sculpt fine details
- Paint and texture the model
- Pose the model
- Release your creation via 3D printing or other export formats

Tap the Done button in the upper-right corner to leave the introductory screens and start modeling your first project.

Find Models in the Sculpt+ Community

Two horizontal menus appear along the top of the Sculpt+ main screen, which is shown in Figure 2-1. The topmost bar is a learning center where you can access lessons on the basics of sculpting. These are very much worth watching.

The bar below contains five options for exploring designs created by members of the 123D Sculpt+ community: Popular, Featured, Recent, My Feed, and Search. The power of this app truly lies in its strong community of users—you can view, share, and comment on the work of your fellow modelers and artists. Even better, you can use their creations as the basis for your own new works—a great shortcut as you get started with the app.

At the bottom of the screen there are three options to select from: Community, Create, and Me.

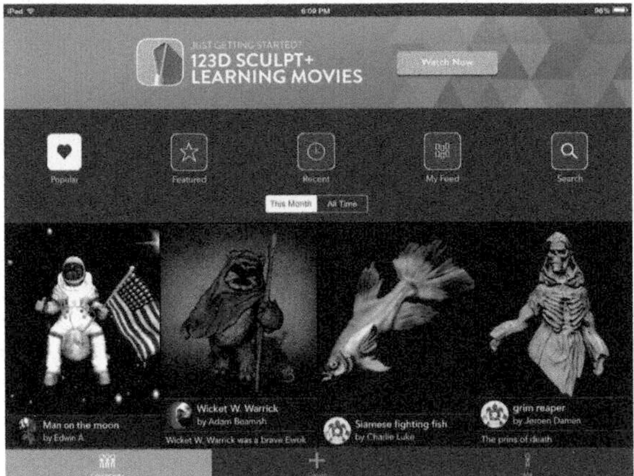

Figure 2-1 *The Sculpt+ main screen*

Tap the Community option on the bottom menu on the Sculpt+ main screen, and you'll find hundreds of models designed by your fellow Sculpt+ users. These amazing models are the coolest things about the Sculpt+ community.

To learn more about a model you like, just tap to select it. A window will appear displaying the model, as well as an info box with the names of the model and its designer, and a brief description (see Figure 2-2). There are options here to leave comments, and "favorite" a model you particularly admire. You can also share it (this option includes a button to flag a model that may be inappropriate or offensive, so that moderators can review it), and best of all, download the model to play with, alter, and reshare.

Another cool thing about the Sculpt+ community is that you can follow users who inspire you, and track their latest efforts.

In the upper-right corner of the model info box, tap the icon shaped like a silhouette with a plus sign, and you'll be notified when this user places a new model in the gallery.

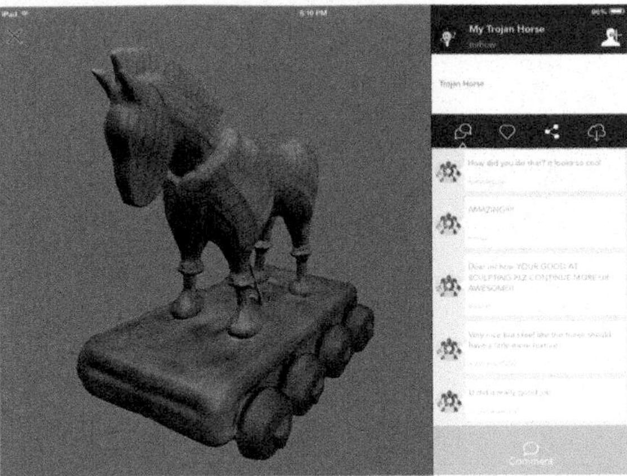

Figure 2-2 *The Sculpt+ model info box*

Tap the big X in the upper-left corner of the model window to return to the main Sculpt+ screen.

Me

The "Me" tab, on the far right of the bottom menu, is the holding area for your own designs, as well as things you've downloaded from the Internet (see Figure 2-3).

Figure 2-3 *The Sculpt+ Me menu*

When saving a model, you have two options: save to community or save to your own account.

The advantage to saving your model to the community is that this preserves your design beyond your iPad. If, for some reason, you delete the 123D Sculpt+ app from your iPad, you can reinstall it, find your models in the community, and download them. But you will lose any models that you saved to your iPad alone.

That said, you may not always wish to share a model you create, so the option to keep them private is also available.

Create

Let's take a look at modeling workspaces available in 123D Sculpt+. To begin, tap the Create option—the giant plus sign—in the middle of the bottom menu. A new screen will appear with three working options (see Figure 2-4).

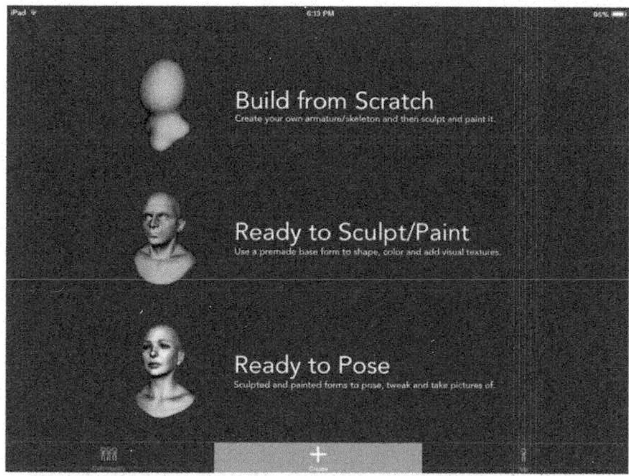

Figure 2-4 *The Create menu*

Build from Scratch

To create a new creature, tap the Build from Scratch option. As shown in Figure 2-5, a selection of basic skeleton forms will appear: biped, head, classic, and so on. Swipe to the right and left to move between these options, and tap one to select it.

Figure 2-5 *The Build from Scratch menu*

Skeleton Workspace

After you've selected a skeleton, you'll move into the skeleton workspace, where you create the foundation for your sculptural model (see Figure 2-6). There are two horizontal menus on the screen. The one across the top contains five options that remain consistent across different workspaces: Menu, Undo, Redo, Help, and Reset Camera. Let's take a look at each of them.

Menu

Tap this to move around the app quickly. For example, tap to go back to the gallery of models; you'll see a prompt that asks if you want to save the changes to your current project.

Undo

A left-pointing arrow. Tap this to quickly undo changes you have made to a skeleton or sculpture.

Redo

A right-pointing arrow. Tap this to redo changes to your creation or re-create features.

Help

A question mark inside a circle. Tap for basic information on how to use the current workspace.

Reset Camera

A cube inside four corner marks. If, as you zoom in on a part of your model, you get lost amid the joints and bones, double-tap Reset Camera. The view will zoom out until you can see the entire model again, and it will return to its original front-facing position.

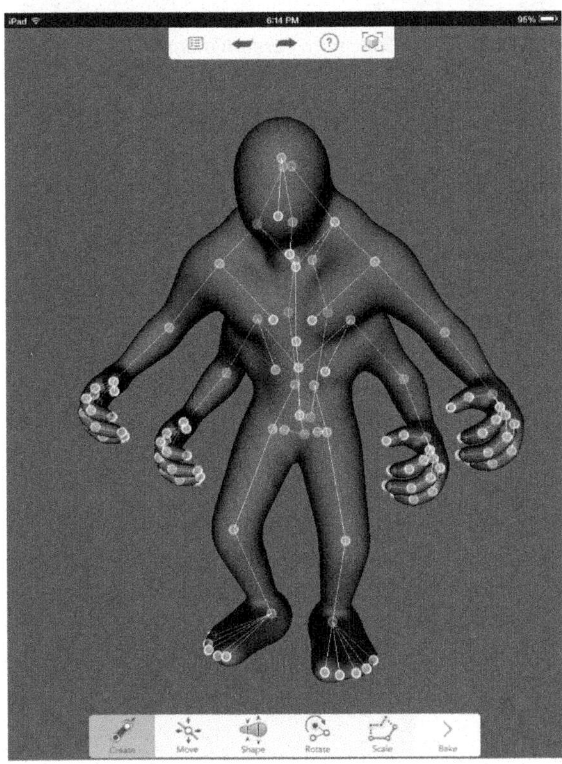

Figure 2-6 *The skeleton workspace*

Be sure to insert as many features as possible in Skeleton mode, so that you'll have enough to work with during sculpting and texturing. Features like teeth, claws, and head shape should be created while in Skeleton mode.

All creations in Skeleton mode are symmetrical.

Navigating in Skeleton Mode

Rotate by tapping the screen outside the model, and dragging your fingertip in the direction you wish the model to rotate.

Zoom

Zoom in by using two fingertips to pinch inward. Zoom out by spreading two fingers apart.

Pan

Pan by dragging to the right or left with two fingertips simultaneously.

Place your skeleton in the view that corresponds with the direction in which you wish to create. For example, you'll want to view your skeleton from a side angle to pull joints frontwards or backwards.

Skeleton Toolkit

To edit in skeleton mode, you use the toolkit that appears along the bottom of the app window (see Figure 2-7).

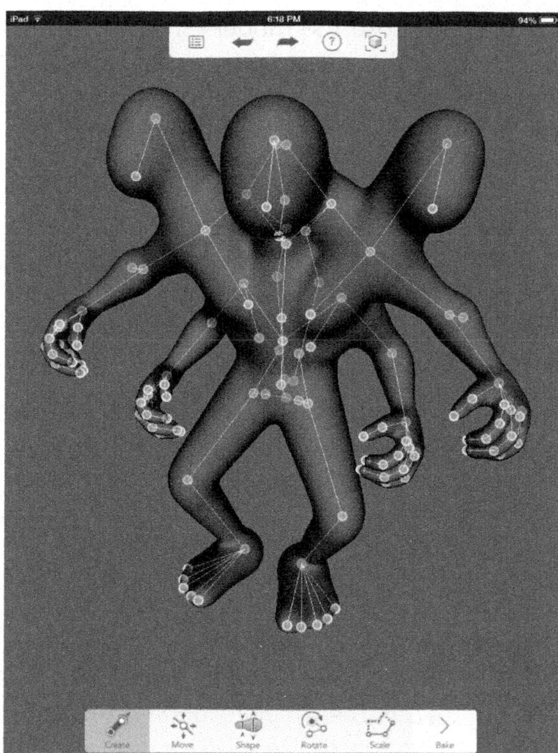

Figure 2-7 *Using the skeleton toolkit*

Create

The Create tool, located at the bottom left of the screen, is what you use to add new segments to your skeleton. Tap Create to select it. Then, touch a preexisting joint with your fingertip and drag in the direction in which you want the new joint to go.

Delete, Cut, Copy, Paste

Select a joint by tapping on it, and a pop-up menu will appear containing icons to delete, cut, copy, and paste.

Add Joint

Tap a bone to add a joint (a dot) to it.

Combine Joints

To reduce two joints to one, grab one and drag it to the other. They will automatically snap together.

Move

Use the move tool (four arrows around a central point) to reposition a joint.

Shape

Shape, the third icon from the left on the bottom toolbar, looks like a little blob with two bands across it. This is a very important tool! Tap to select. Then, on your skeleton, drag the bands up and down to thicken or thin a bone (Figure 2-8). This is a great way to create head, body, leg, or arm shapes.

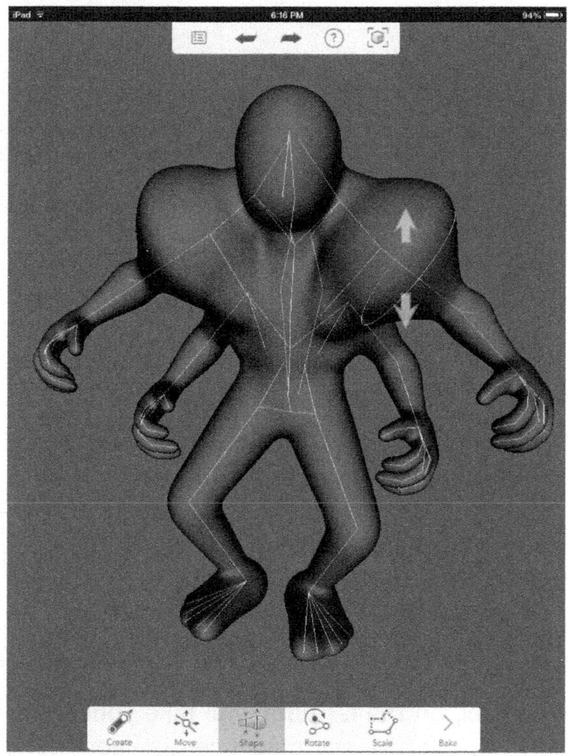

Figure 2-8 *Using the shape tool*

Rotate

To move an entire limb, tap the Rotate icon, third from the right, which looks like three joints connecting two limbs. Then select a joint on the limb you wish to move and drag your fingertip; the entire limb will change position as seen in Figure 2-9.

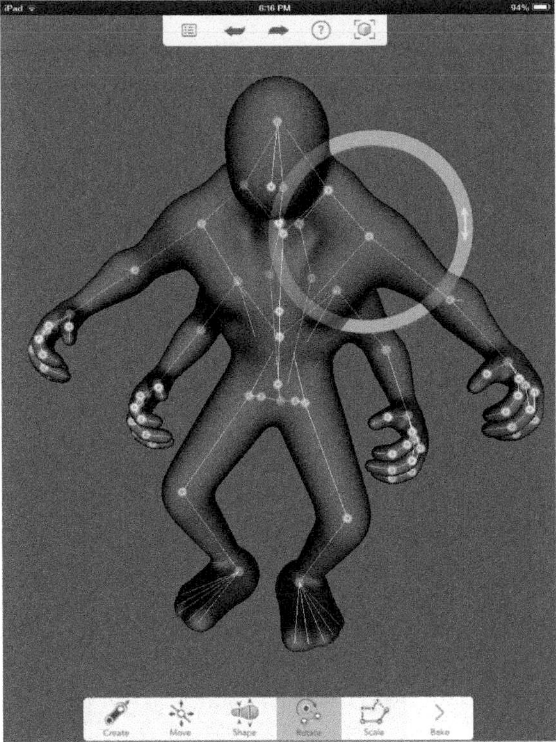

Figure 2-9 *The Rotate menu*

Scale

This is the second icon from the right, a six-sided shape. Tap it, then drag a joint to scale a limb or series of limbs. This is a great tool for adding hand details, because you can create a detail in any size you want, and then use this tool to scale the entire hand down.

Bake

This tool, on the far right, will take your skeleton through the evil mad scientist laboratory, turn it into a model, and deliver it to the sculpting workspace. *You cannot go back.* You can open as a skeleton again but you will lose any sculpting that you have done.

Sculpting and Painting Workspace

This workspace is the place to add detailed features and textures to the model, add paint and texture, and pose the model—in other words, this is where the magic happens. You can even add an image to put your model into an environment, and then release your creation via 3D printing or other export formats.

Sculpt

Tap to select the Sculpt tool on the bottom right, and a new menu of options will appear that allow you to create detailed features on your model.

Each option can be further refined with the two sliders that appear above them. The one on the left controls the tool tip size, while the slider on the right (with a little lightning bolt icon alongside it) adjusts the tool's degree of force.

Start with the Grab tool: use it to move things into the places you wish. Then, before getting into the finger details of the face or body, set the Sculpt Out tool with a larger brush and mid-level strength, and use it to build up muscle on the model.

Back

Tap this any time to go back to the main sculpting workspace.

Sculpt Out

With Sculpt Out, shown in Figure 2-10, you can set your brush size and strength, then drag your fingertip to bulk out an area of the model. You can use this to build up muscles, nose, brow, cheekbones, chin, and ears.

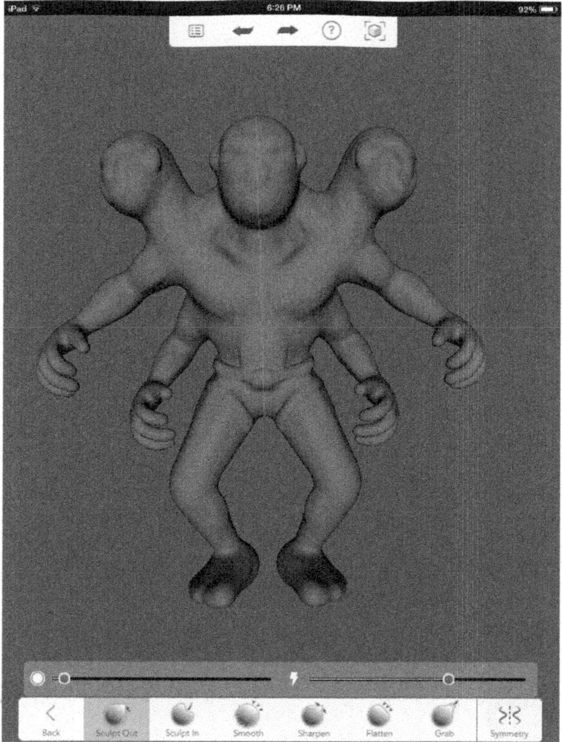

Figure 2-10 The Sculpt Out menu

Sculpt In

With Sculpt In, shown in Figure 2-11 and Figure 2-12, you can tap and drag to carve more detail into the parts you sculpted out. This is great for ears and nostrils.

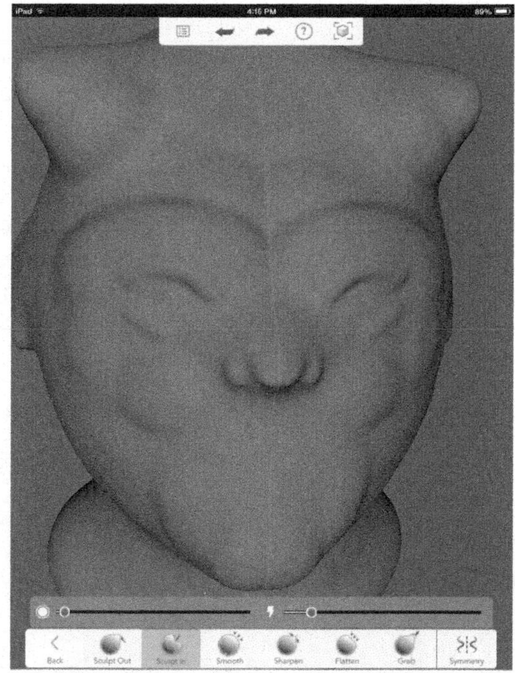

Figure 2-11 Sculpt In allows for more detail...

Figure 2-12 ...especially in body parts

Smooth

Drag with your fingertip to soften and round areas of the model. This can be used to fix small mistakes and pixelation.

Sharpen

A great tool for defining details such as eyelids or the bridge of the nose. Make small circles, or tap and drag to create a sharpened area. This tool is a bit more subtle but can really help you get clean edges (Figure 2-13).

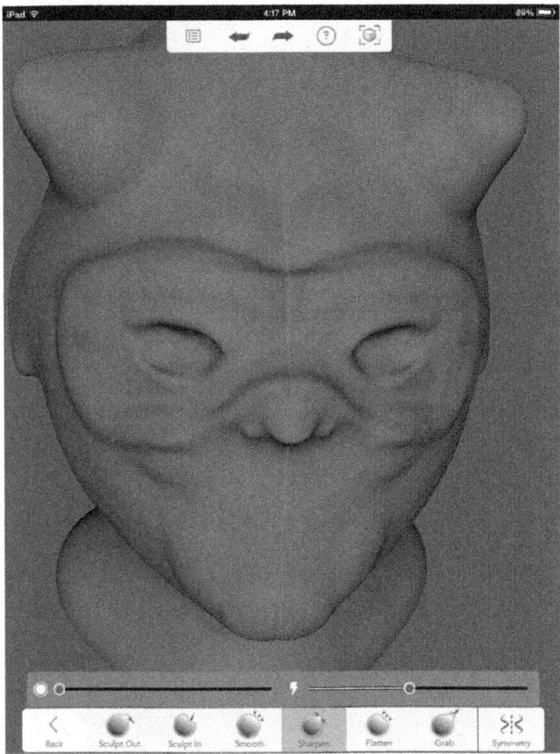

Figure 2-13 *The Sharpen tool*

Flatten

This tool can be used to add a hard edge to a flattened area. Use it to sculpt the angles of the face, or to create a defined chest or abs. It is also useful for creating clothing (like hoods, see Figure 2-14) and accessories.

Figure 2-14 *Use Flatten to create realistic looking cloth*

Grab

The Grab tool is used to stretch a feature. With a large brush size and mid to high strength, the Grab tool can move limbs around to make them less symmetrical. It can also be used to pull things out of the head shape, like ears, horns, or wings—although it is highly recommended to place those on the skeleton before baking it.

Grab is a powerful tool: for example, take a look at Figure 2-15 and Figure 2-16, which show the before and after, respectively, of the ready-to-sculpt bust of a man. The features were created using only the Grab tool.

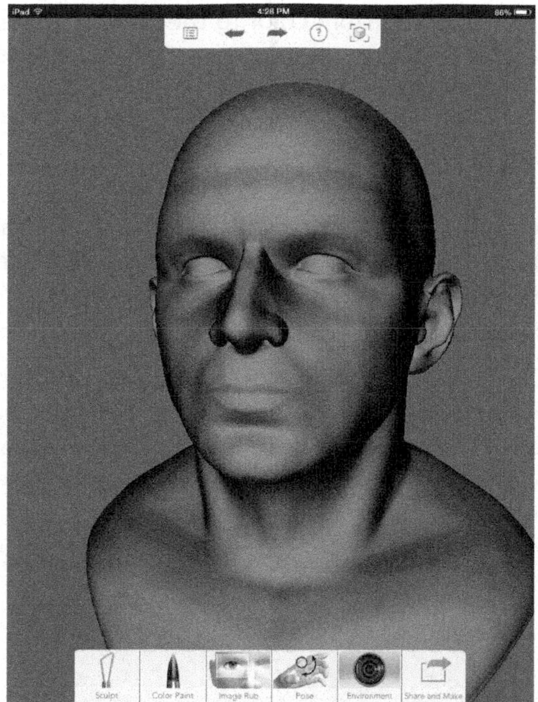

Figure 2-15 *Before using Grab*

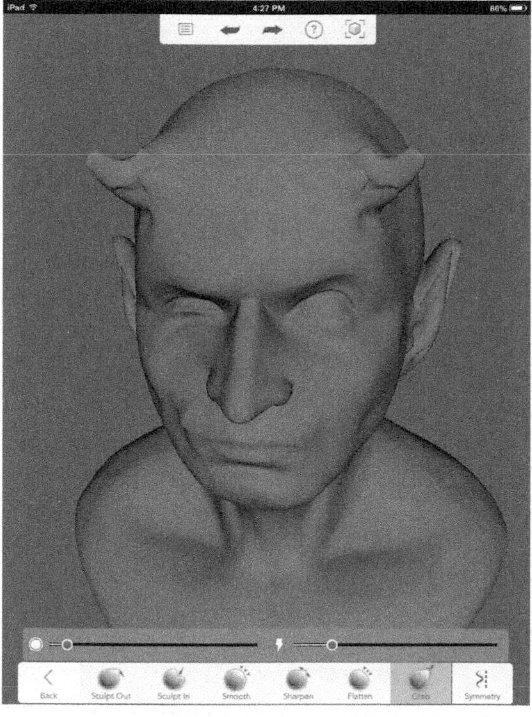

Figure 2-16 *After using Grab*

Symmetry

Toggle symmetry on and off by tapping the button in the lower-right corner of the design workspace. This allows you to sculpt one feature independently of its twin. This can be a great tool to use while building out the model's initial shape.

Now it's time to paint!

Color Paint Toolkit

Painting brings your model alive with color and texture (Figure 2-17). It's also a means to cover up some of the areas that may not be perfectly formed on your model. You use various colors of digital "paint," applied with a variety of digital "brushes," to color your creation. Always try to paint slowly and build up layers and texture. This usually means keeping your brush to a smallish size (though not tiny), and your strength down.

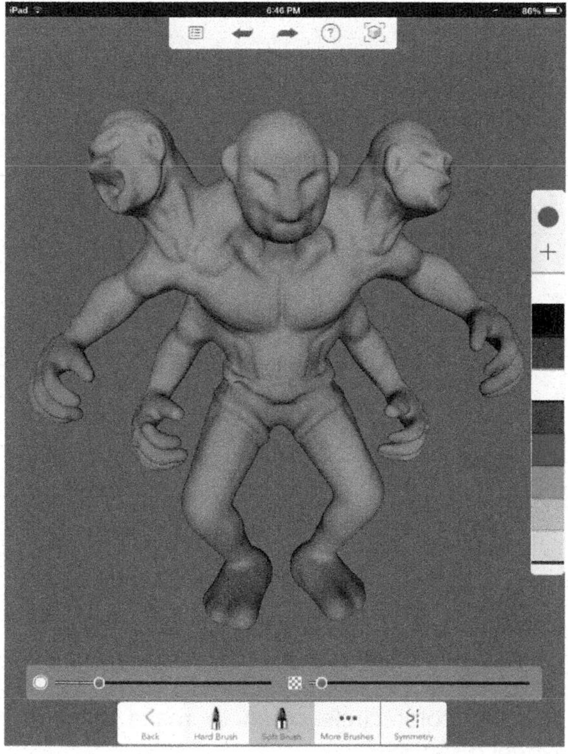

Figure 2-17 *Color your models with Color Paint*

Color Paint

Tap to select this tool set, and a menu of brushes appears along the bottom of the screen, while color swatches appear on the right.

Color Swatch Options

Change your paint color by tapping on the color you wish to use (squares), and the selected color will appear at the top of the swatch menu (circle); see Figure 2-18. To create custom swatches, tap the selected color to bring up a color wheel. Adjust the different settings in the color wheel until you find one you like. Tap the inside of the diamond to change the saturation of the color. Tap and drag the outside circle to select the color.

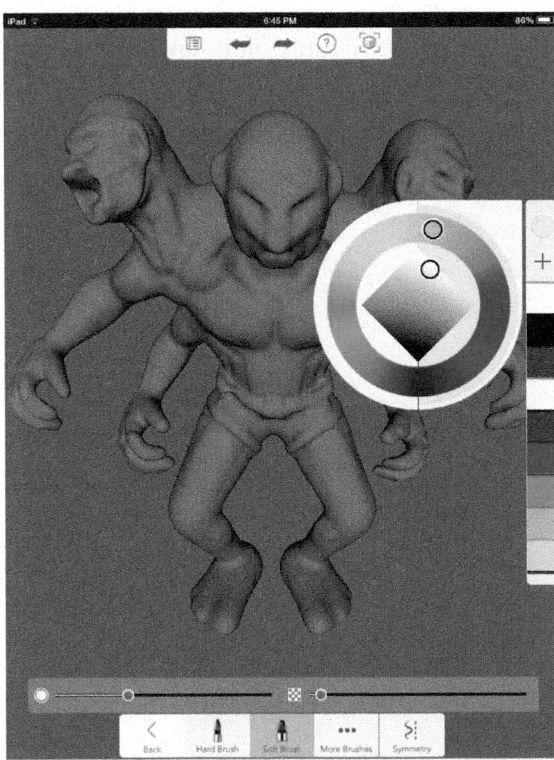

Figure 2-18 *Using the Sculpt paint color picker*

Tap the color circle again to make the color wheel disappear.

To add your custom color to the swatch menu, tap the plus sign that appears right below the selected color.

Brush Options

As with the Sculpting tool set, you can control the brush size with the slider on the left, while the one on the right controls opacity/transparency (Figure 2-19).

On the right side of the bottom menu, you can toggle symmetry on and off to add a color or texture to one limb or side of the model without the same effects appearing on the other.

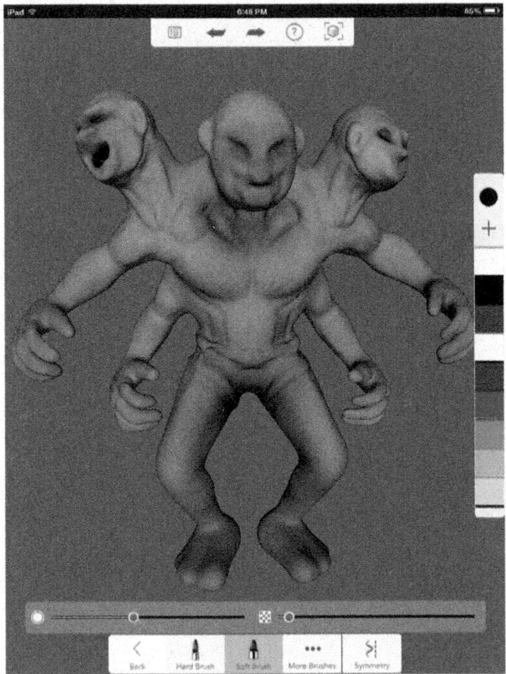

Figure 2-19 *Some of the Brush options available in Sculpt+*

Apply a base coat to your model, making sure that symmetry is toggled on.

Hard Brush

Brush across your model to add color. Use a lower strength setting to build up color or texture slowly.

Soft Brush

This extremely useful tool creates soft, blurred edges, which is great for blending colors together—it comes in handy for creating realistic skin tones, for example.

More Brushes

Tap this option to discover a number of different paint effects: splatters, streaks, and so on.

Image Rub Toolkit

Tap Image Rub to bring up the tools that let you apply a pattern or texture to a model's surface. To use this tool, you select an image that represents the texture you want to apply to the model. Then you select the brush you wish to use, adjust the brush size and opacity, and then apply the brush to the model to add the texture (Figure 2-20).

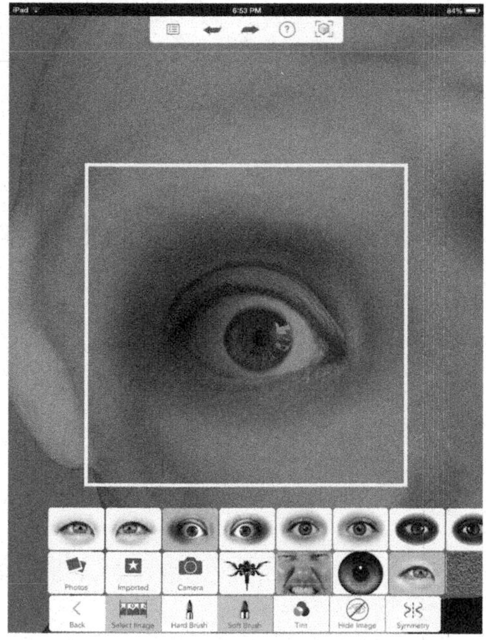

Figure 2-20 *Applying a pattern with Image Rub*

You may have to reposition the model to add texture to its entire surface.

Hard Brush, Soft Brush, and Symmetry work here much as they do in the Color Paint toolkit.

Tint works identically to the custom swatch menu in the Color Paint toolkit.

Select Image

Tap this tool to select an image to apply to the surface of your model. Select Image contains several options. Let's take a look.

Photos

Allows you to add a photo from your tablet's Photo Album, which will then be accessible under the Imported option. This is a great option for adding certain textures or patterns. We use this all the time. Figure 2-21 shows a great example from user Kylegarvin.

Figure 2-21 *Adding a pattern from a photograph*

Camera

Uses your tablet's camera to take a new picture to use as an image, without leaving 123D Sculpt+.

Use your camera to create background photographs, bring them into the workspace, and place your model into a lifelike scene.

Pattern

Displays preloaded line drawings that you can rub on your model, which is great for creating patterns on clothing. This option can sometimes also be used to create tattoos.

Face

Brings up a number of preloaded photos that make it easy to give your model a realistic expression.

Eye-Related Icons

Brings up, you guessed it, a selection of iris and eye shape options, which are incredibly useful for detailing your model's eyes.

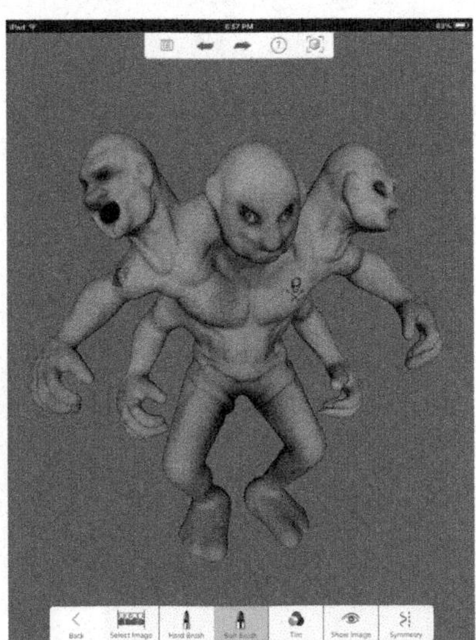

Figure 2-22 *Creating highly defined eyes and irises, for truly creepy/realistic results*

Mottled Gray Texture Icon

Brings up a selection of surface textures that help create satisfying textile, skin, leather, or armor textures. For example, check out the texture shown in Figure 2-23.

Figure 2-23 *Creating a mottled, pebbly texture on your model*

Three Stars

Contains a selection of tattoos that you can apply to your model (see Figure 2-24).

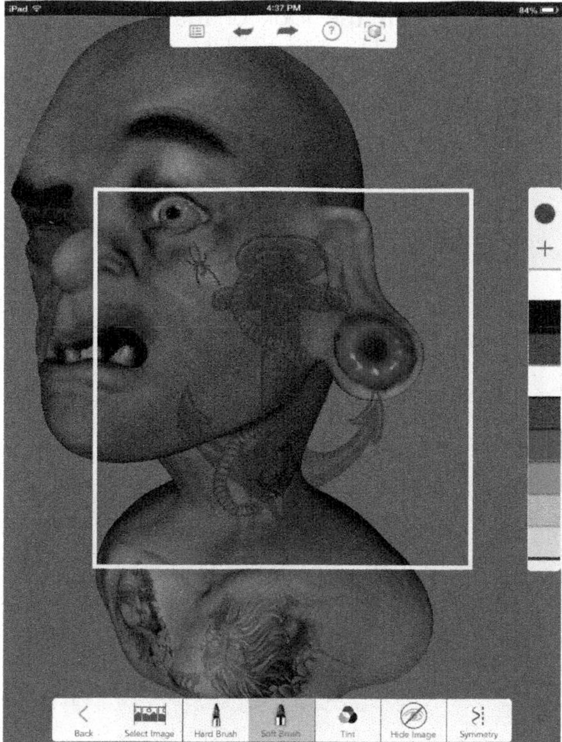

Figure 2-24 *What good is a monster without tattoos?*

Hide Image

Tap the Hide Image option to hide the master image overlaid over your model, so that you can see your model more clearly. You can continue to apply the image while it's hidden, or tap Show Image to bring back the master overlay.

Pose Toolkit

Realistic 3D creatures shouldn't spend all their time standing at attention. Placing their limbs and bodies into human-like (or animal-like, or even otherworldly) poses will make them seem to come alive. The Pose toolkit offers two methods to place your creature into more natural postures.

Pose

Tap to select, and then double-tap on your model to activate joints and position limbs and other appendages. Touch the screen outside your model and drag in order to reorient the entire model—important if you want to position an appendage in a particular direction. This is shown in Figure 2-25 and Figure 2-26.

Figure 2-25 *Double-tap the model to activate joints...*

Figure 2-26 *...and drag to reorient the limbs*

Orient

This offers many different lighting options for users who wish to light their model in a unique way in order to enhance a few features.

Environment Toolkit

The Environment toolkit allows you to adjust the look of the space around your model, so that it looks much better and more engaging than a plain gray screen. It also contains some options for giving your model a polished and professional final appearance.

Use the Background button to add a backdrop image behind your model; you can select from a variety of sources, including your tablet's photo album, or even create a new background image on the spot with the tablet's built-in camera. Tap Lighting, and choose from nearly a dozen preset angles and intensities. Use Effects to quickly shift the look of the entire scene, both model and background (Figure 2-27).

Figure 2-27 *The Environment Toolkit adjusts the environment around your model*

Share and Make Toolkit

Finally, the Share Image tool is the one-stop shop for sharing your final design—model, environment, and all—via Facebook, Twitter, email, instant message, and other channels that you may have added to your tablet. See Figure 2-28.

Figure 2-28 *And when your model is done, share it all around!*

Once you have finished your model, tap Share and Make. You can publish your creature to the community, export your finished creature as a mesh to other programs for polishing or printing, or create an order for a 3D print via Sculpteo, a third-party provider.

123D Meshmixer

3

Meshmixer is the most powerful program in the 123D family. Using Meshmixer, you can merge parametric models with surface models to create your own mashups and mix-ups such as these Open Cell Bunnies (Figure 3-1).

Meshmixer also offers up a full slate of features, including sculpting, stamping out shapes, painting, creating surface patterns, running several types of analyses, and a utility that assists you in creating watertight models that are ideal for 3D printing.

You can also use Meshmixer to print directly to your desktop 3D printer, or send your design to third-party printing companies such as i.materialise, Sculpteo, and Shapeways, which offer diverse options for materials, colors, and finishes.

This overview will try to give a good taste of some of the more popular and spectacular uses of Meshmixer.

Figure 3-1 *Image by Christian Pramuk-Autodesk*

Why Use Meshmixer?

You can do a slew of things with this program, but these are the two you're likely to do most often:

- Merge surface files, called *.stl* files, into a single part that can be 3D-printed.
- Create highly intricate and organic shapes.

Getting Started

When you first launch Meshmixer, the home screen appears, presenting a series of options and essential tools in the middle of the window (Figure 3-2).

Some tools also appear in a menu down the left edge of the window, where they can be accessed when a project file is open.

The following sections take a look at what these tools do.

Import

Imports any *.stl* or *.obj* file.

Open

Opens an existing Meshmixer file.

123D

This button, which has a sphere on it, opens up the 123D library, where you can find and import parts into Meshmixer.

Import Bunny

A magical whistle that brings a bunny object hopping onto your screen, where you can play with it.

This bunny is famous among 3D-graphics nerds. It's a 3D test model containing 69,451 triangles, developed in 1994 by Marc Levoy and Greg Turk at Stanford University. The Stanford bunny, as it's popularly known, is available for free download, and is one of the most frequently used models out there. Over the years, we've seen thousands of variations on the basic Stanford bunny, from small to large and simple to complex.

You can read more about this model on Greg Turk's bunny web page, "The Stanford Bunny" (*http://bit.ly/stanfordbunny*).

Import Plane

In this case, we are referring to a flat rectangle, not the "main plane" workspace of 123D Design and other CAD programs.

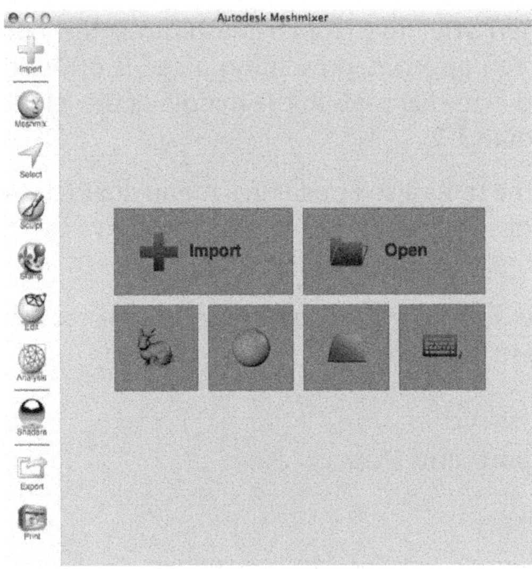

Figure 3-2 *The Meshmixer home screen*

Keyboard Shortcuts

Opens a PDF file containing a three-page visual reference (Figures 3-3, 3-4, and 3-5) for Meshmixer's load of hidden keyboard and mouse commands. You don't need to know any of them to use the program, but they can save you keystrokes and speed up your work.

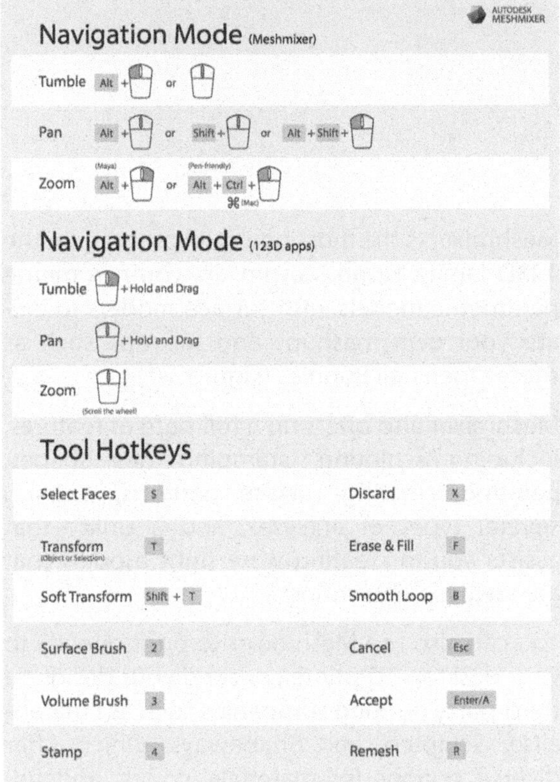

Figure 3-3 *Keyboard shortcuts for 123D Meshmixer*

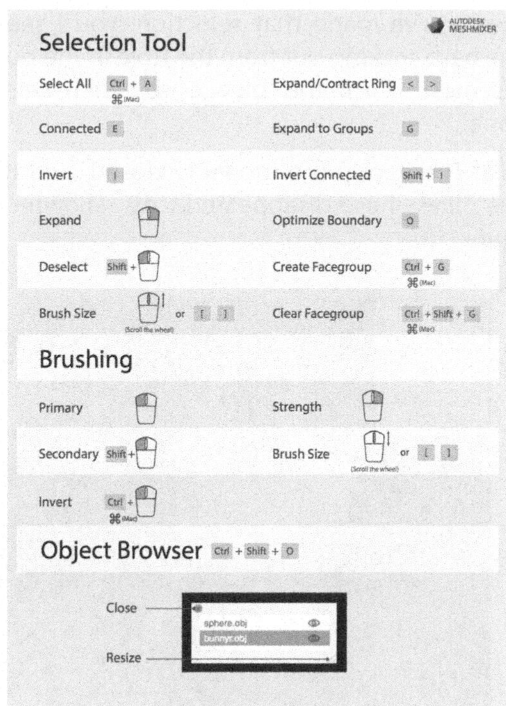

Figure 3-4 *More 123D Meshmixer keyboard shortcuts*

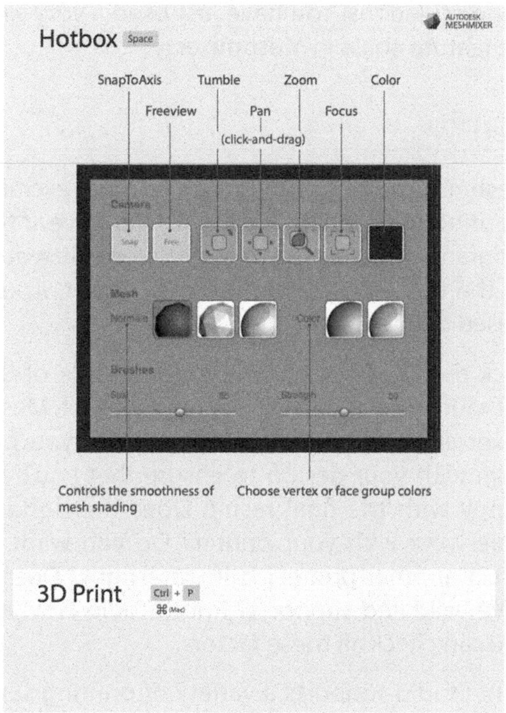

Figure 3-5 *And more 123D Meshmixer keyboard shortcuts*

Tutorial: Seal Up the Bunny

In this tutorial, we open up a sample model, and then fix a gap in the design so that it is watertight. A model must be watertight in order to be printed in 3D.

When a model is watertight:

- Every triangle edge has exactly two neighbors, which means there are no holes or "nonmanifold" edges—that is, a mesh edge with more than two entities attached to it.
- Every node in the triangle is connected to only one "fan" of triangles around it. If the model is watertight, every triangle that shares a node is accessible from any other triangle in the same node by moving across the triangle edges.
- No triangles overlap or intersect with each other geometrically.
- No geometric errors occur that result in unrealistically thin areas in the model.

At the time of this writing, Autodesk provides some pointers on ferreting out watertight problems with models on this web page (*http://bit.ly/watertightproblems*).

Step 1: Import the Bunny

When you first open the program, you will see the options to import, open, or use the 123D library. Under that, you will see the shapes of a bunny, a sphere, a plane, and a keyboard. Click on the bunny. (Just a heads up: we will be using the same bunny throughout this chapter.)

Select "import bunny" (Figure 3-6).

Figure 3-6 *Import the bunny prior to rotation*

Step 2: Find the Gap

Notice that there is a blue line around the bottom of the bunny. That blue line indicates that there is a gap somewhere in the object, which prevents it from being watertight.

Rotate the object to see the opening by holding down the right mouse button and moving the mouse, until you can see up inside it (Figure 3-7).

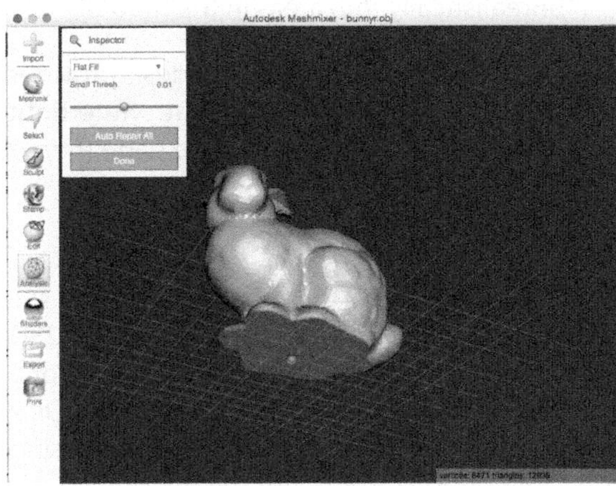

Figure 3-7 *The underside of the bunny*

Step 3: Close the Gap

Click the Analysis tool, which you'll find midway down on the toolbar.

Select the Inspector option on the top of the menu.

Once you've made that selection, you'll see a blue barb sticking out from the hole in the bunny's mesh. Click the blue barb. Then select "done."

Voilà! The gap in the model is closed, and the blue line has disappeared, as shown in Figure 3-8.

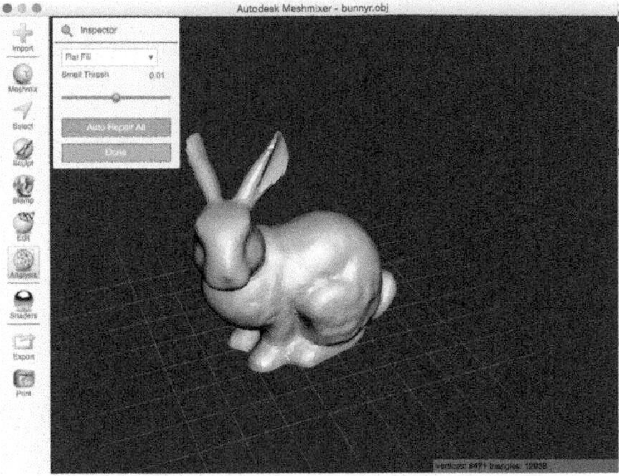

Figure 3-8 *The blue line vanishes when the gap is closed*

Congratulations! You have just used a very simple feature chain in Meshmixer.

Print

Meshmixer's print function is the most exciting 3D printing experience you will ever have. If the program consisted of this tool alone, it would be the best 3D program on the market, in our biased opinion.

Click the Print button on the lower side of the lefthand menu, and a separate area of Meshmixer appears: the Print Studio. Here you can work with your design to ensure that you'll be happy with the final result. Does the design's scale work with your printer? Do you want to select another printer? Is it watertight? Does it have validated supports? Print Studio is where you can check all these factors.

Print Studio supports a variety of printing services, as well as 3D printers. You can select the service or printer you want to use by clicking on the pre-selected printer, right below the the Import, Open, and Modify buttons. If you do not see the printer on the list of options, select Choose Printers (Figure 3-9) and a new window will pop open. Select the printer you wish to use from those on the available list.

If your printer isn't listed, click the Add New Printer button towards the bottom right of the window.

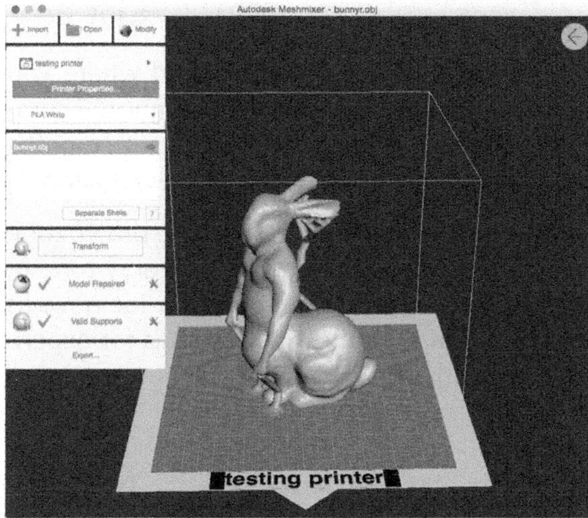

Figure 3-9 *Choosing the printer*

The Printer Properties window pops open. Here you can enter a name for your new printer, as well as the Printer Parameters: the units of measurement, maximum printing dimensions, and so on. If you are uncertain what these settings need to be for your new printer, check in the printer instruction manual, or with the machine manufacturer's website. Most 3D printer makers post this information online.

From here, you can select to repair the model and then select to validate the supports (if you don't have supports, it will generate them for you).

Then select Export to get your watertight, supported STL and open it up in your printer driver.

Now go back into Meshmixer by clicking on the arrow in the upper-right corner of the workspace.

Toolbar

Now let's investigate the other tools in the vertical toolbar running down the left side of the window.

Import

Imports additional or new models into the workspace.

When importing you can choose to append (add on to) or replace the model that is currently in your workspace.

Meshmix

Offers a variety of simple models that can be added to your model, such as a torus, cylinder, or cube. There are several menus to choose from; if you click the Primitives option, you will find fun things like arms, ears, heads, numbers, and our favorite, miscellaneous.

To add one of these parts to your model, click and drag it from this menu to the desired area on your model.

Let's try one.

Step 1: Import the Bunny

If you did not already import the bunny in "Tutorial: Seal Up the Bunny", then go to File → Import, and import the bunny (Figure 3-10).

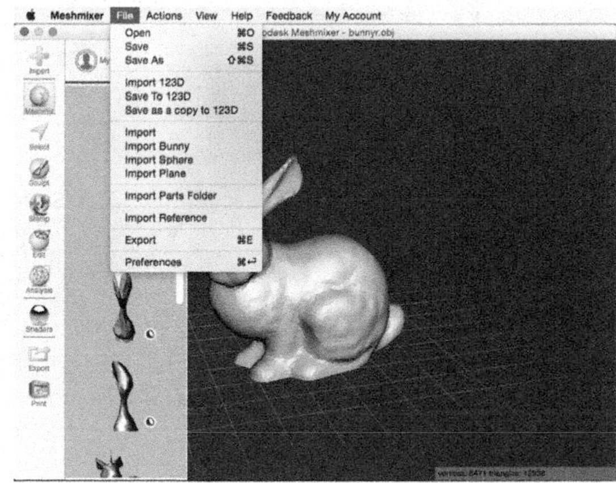

Figure 3-10 *Import the bunny*

Step 2: Heads Library

Select Meshmix from the menu options, and find the Heads library (Figure 3-11).

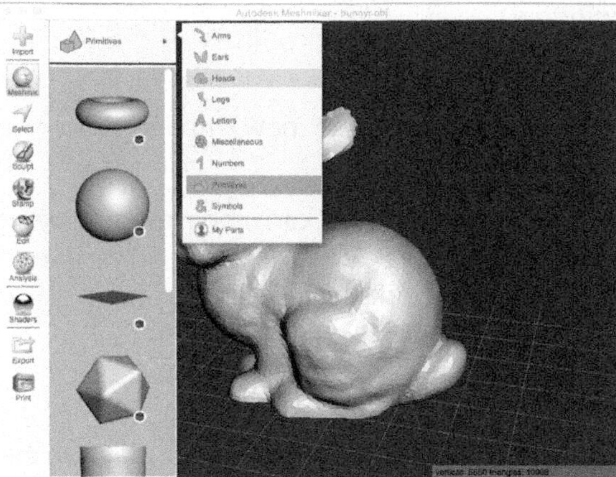

Figure 3-11 *Meshmixer offers access to a library of preformed head designs.*

Step 3: Select and Merge the Bear Head

Grab the bear head from the available options, and pull it over to the bunny's butt.

Toggle your model by clicking and holding down your right mouse button until you can get a good view of the bear head. Zoom in by rolling your middle mouse wheel.

Use the toggle wheel under the bear head to move it around and resize it (Figure 3-12).

Once you've placed the bear head where you want it on the bunny, toggle the SmoothR command up and down, and watch as it smooths out the rough edge between the two objects.

Click Accept, and you'll see that the brown tone of the bear's head changes to the gray tone of the bunny. Keep this bunny open for now, as we'll use it in the following sections for experimenting with other tools.

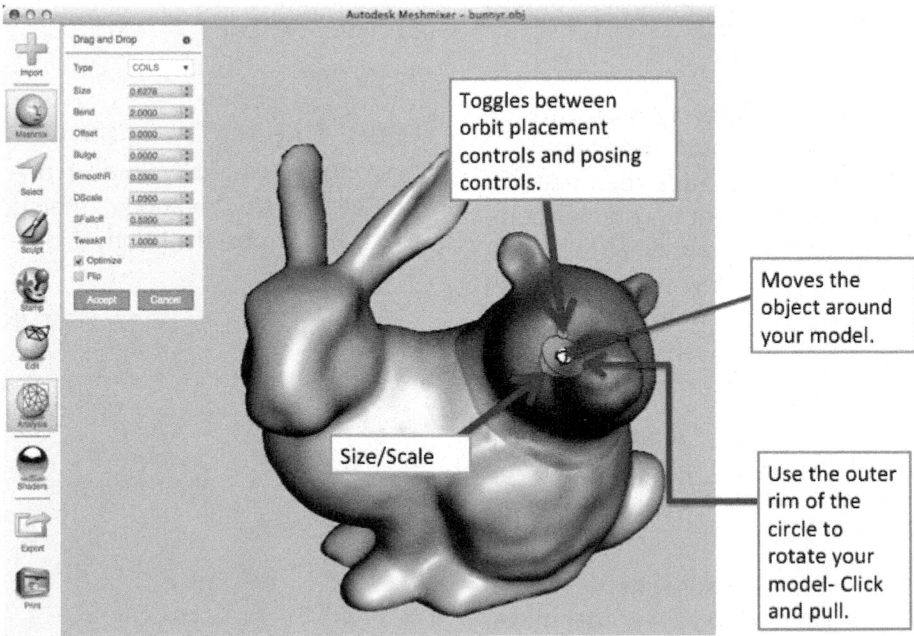

Figure 3-12 *Use the toggle wheel to move the bear head around*

My Parts Library

You can also create your own library of parts in Meshmixer. This can be very useful if you are building several objects that utilize the same components, such as stackable plastic bricks or other interlocking parts.

To add a part to your part library:

1. Have the part you wish to add to your library open in your workspace.
2. Press Command-A on your keyboard to select all. You will automatically be placed in the select menu.
3. Go to "Convert to..." and select either Open Part or Solid Part. This will depend on if there is a hole in your part (e.g., as in a vase) or if your part is closed (e.g., as in the bear-butt bunny we're creating here).
4. Go to Meshmix and check in the My Parts folder (Figure 3-13).

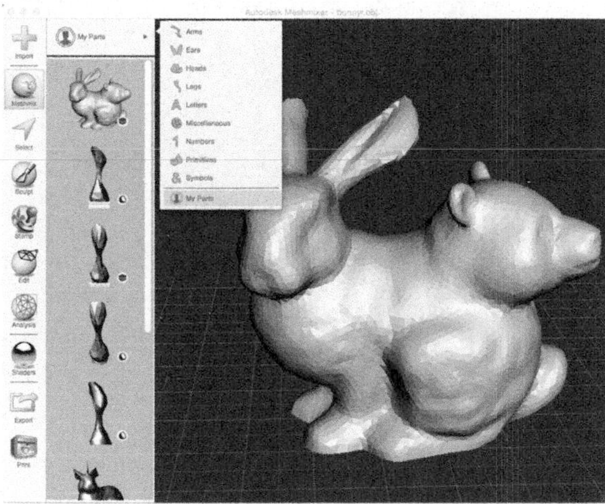

Figure 3-13 *You can add the bear-butt bunny to your My Parts library for future reuse*

Select

The Select tool allows for various ways of selecting and altering different parts of your model (Figure 3-14).

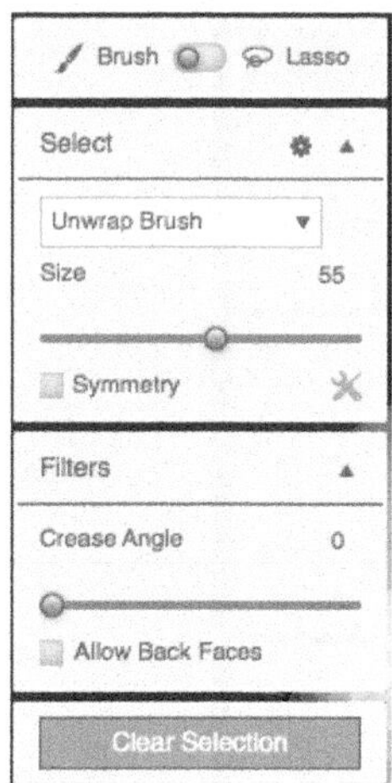

Figure 3-14 *The Select menu*

1. With the top button switched toward Brush, you can paint on your selection. Try painting a nose on the bunny.
2. From the Edit menu that pops up, select Extrude. Play around with the bars and toggles to see what happens. Then click the Cancel button (Figure 3-15).

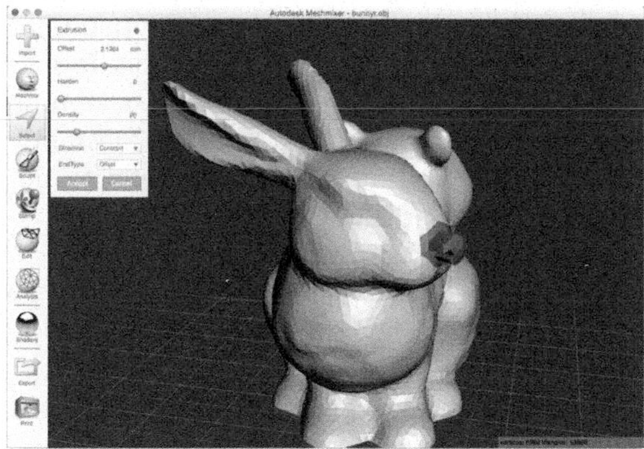

Figure 3-15 *Select offers different options for detailing your model*

3. Play around with the other choices under the Edit menu to see how they work, especially Discard, Extract, and Offset (Figure 3-16).

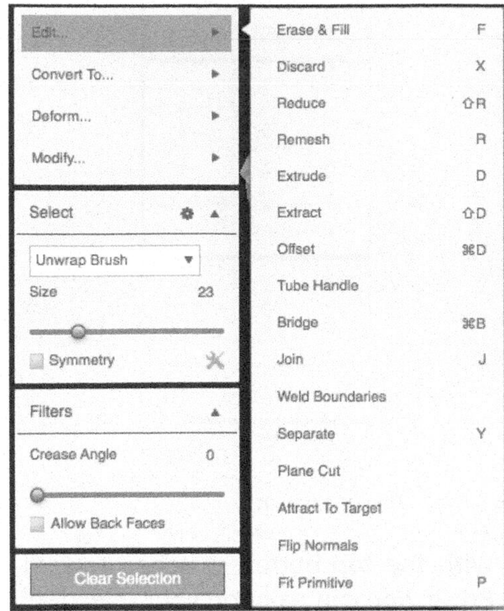

Figure 3-16 *Additional options under the Edit menu*

Sculpt

The sculpt tools include a number of different brushes, each with its own settings, that allow you to draw and carve details into your object. Explore these toolkits, which offer options to create remarkably fine detailing. This tool is simple when you begin but can take a lifetime to master. Just a tip: you should constantly change your strength versus size ratio to get more controlled results.

Volume Versus Surface

When you select the Sculpt icon off the left side menu, you will note that on the top you can toggle between Volume and Surface. What is the difference?

Volume controls the physical shape of the object while Surface controls only the outer skin of the model. These do overlap a bit depending on what brush you use. One way to look at these options is that Volume is plastic surgery while Surface is tattoos.

Edit

The Edit tools are the best way to make big shape edits to your file. You can use them to mirror, transform, and separate out your object. You can also separate your larger models into smaller sections if you want to go big by using the Make Slices tool.

Let's use the Edit tool to turn the outside shell of the bunny into a cool lattice structure, as depicted in Figure 3-17. This project was created by Instructables member Marshall Peck.

Figure 3-17 *Using Edit to make a lattice structure*

Step 1: Import Model and Reduce Polygons

1. Again, bring the bunny into Meshmixer by going to File → Import Bunny. Heal the model again by going to Analysis → Inspector → Auto repair all.

2. Select the entire model using Ctrl-A on your keyboard.
3. Under the Select tool menu, go to Reduce.
4. Increase the percentage slider to about 90%. You can play around with this, but try to get to the point where you see big triangles without distorting the shape (Figure 3-18).

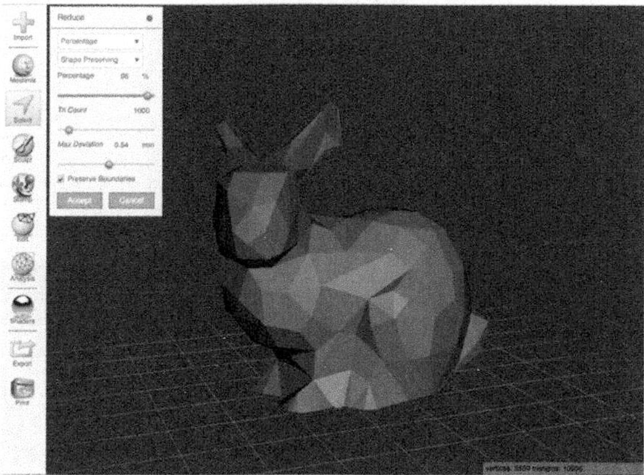

Figure 3-18 *Blocky bunny*

5. Click Accept.

Step 2: Add the Pattern

1. Under the Edit menu select Make Pattern.
2. In the first drop-down menu change Tiled Tubes to Dual Edges. Mesh + Dual Edges generates a pattern inside of the model, which makes a much more complex looking model, as seen in Figure 3-19 and Figure 3-20. Changing the Element Dimensions slider makes thicker or narrower tubes. We have moved ours to 1.257, but if you are printing on a MakerBot-style filament printer you may want to go bigger, otherwise you may have printer problems.

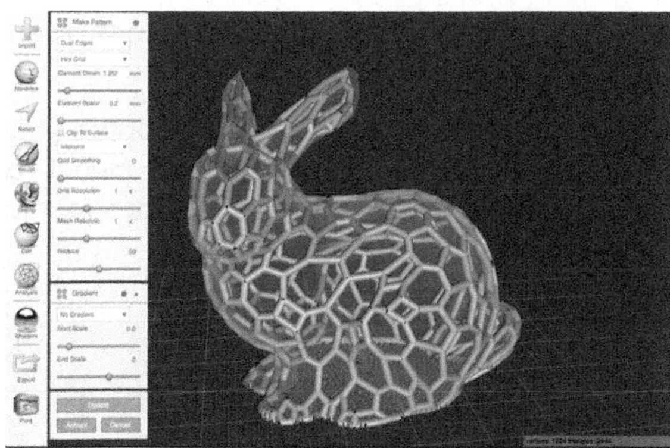

Figure 3-19 *Edge bunny*

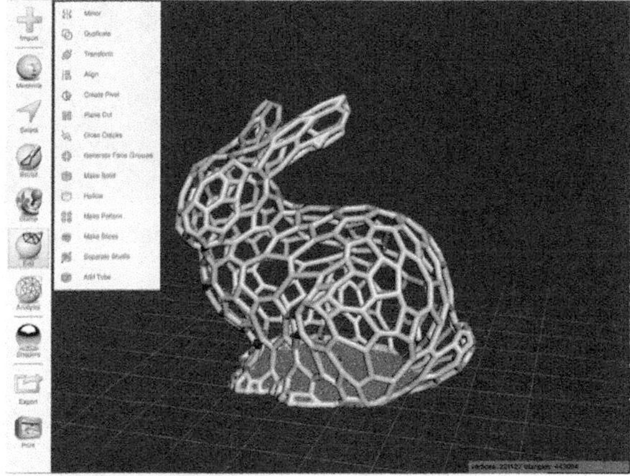

Figure 3-20 *Click Accept*

3. To save the model, go to File → Export .STL.

Analysis

This is the most important menu inside of Meshmixer. It should be used on every model that you ever want to fabricate with a 3D printer.

We have already used this tool menu a few times in this chapter. In Analysis, you can Inspect all of your models to ensure they have no holes in them that would stop them from printing. You can bring in models from any program. You can also generate support for any areas that have a large overhang that your printer may not be able to create, as shown in Figure 3-21 and Figure 3-22.

To do this, go to Analysis → Overhangs. Then click Generate Support.

If you see areas that need more support, you can always click and drag from your model to the grid. That will generate an arm of support for you.

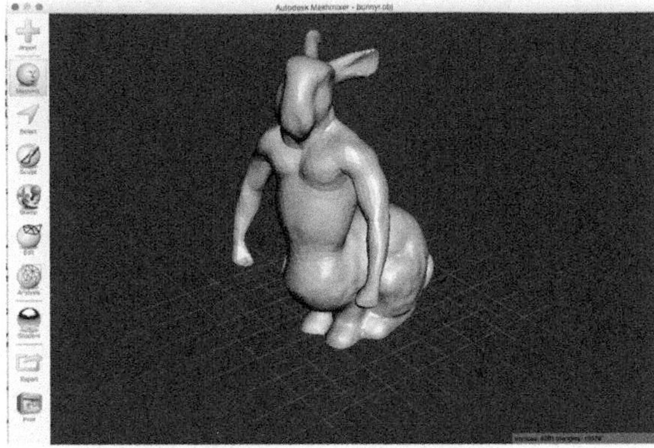

Figure 3-21 *If your model has overhangs that need support during the printing process...*

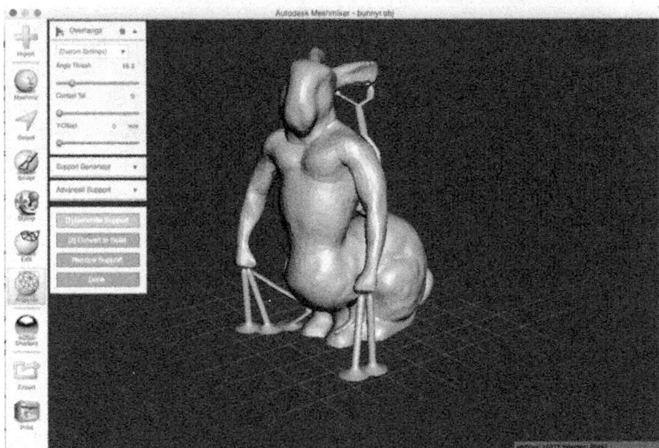

Figure 3-22 *...you can generate support for them with just one click*

Inspiring Meshmixer Projects

The 3D community has used Meshmixer to create some amazingly organic-looking sculptures. As you can see from the diversity of objects shown in Figure 3-23 through Figure 3-28, the only real limitation is in your mind.

Figure 3-23 *A whale by Jesse Harrington Au*

Figure 3-24 *An example of the Make Slices tool by Christian Pramuk*

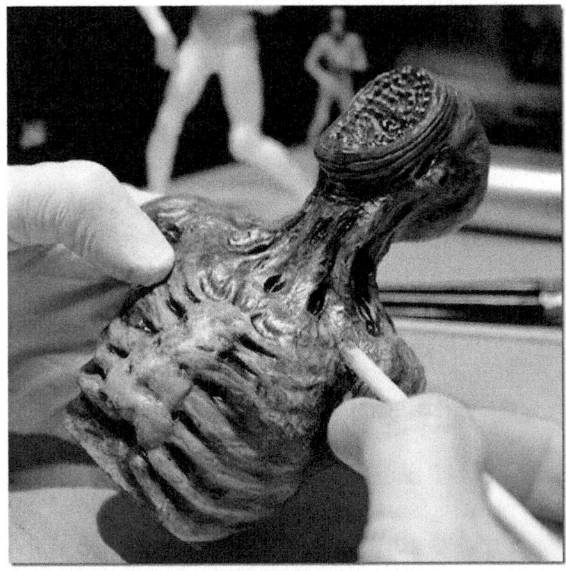

Figure 3-25 *By Craig Barr*

Figure 3-26 *By Gian Villamil*

Figure 3-27 *By Instructables member 3D Printing Ninja*

Figure 3-28 *By Instructables member Le Fab Shop*

You can learn more by checking out Autodesk 123D's YouTube channel (*https://www.youtube.com/user/123d/*) and looking for the Meshmixer 101 playlist (*https://www.youtube.com/user/123d/playlists*).

4 123D Design

123D Design is computer-aided design software that allows you to create precise, even complicated parts for models and objects of all sizes. This program totally rocks at mechanical designs from small to house-sized, such as a housing for your Raspberry Pi (Figure 4-1) or a prototype of your next playground structure (Figure 4-2).

Figure 4-1 *A model for a Raspberry Pi enclosure*

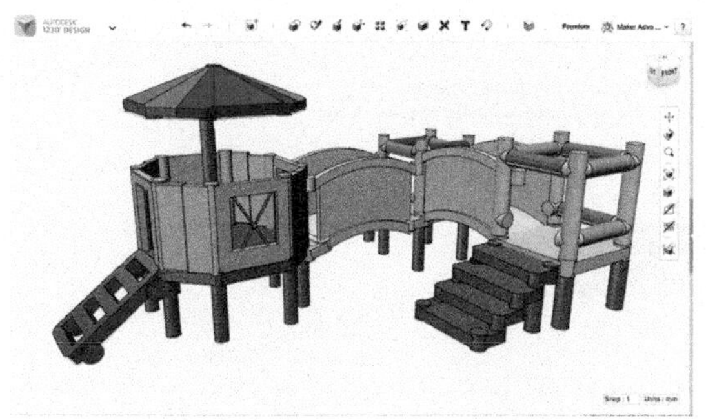

Figure 4-2 *A model for a backyard play structure*

You can import any *.stl* or *.obj* file into 123D Design. You also have access to thousands of existing models in the 123D online gallery, which you can access via the Open File command as long as you are connected to the Internet at the time.

Figure 4-3 and Figure 4-4 show some killer examples of things that users have built with 123D Design, from quadcopter housings to Iron Man's armor.

Figure 4-3 *Some examples of models created with 123D Design*

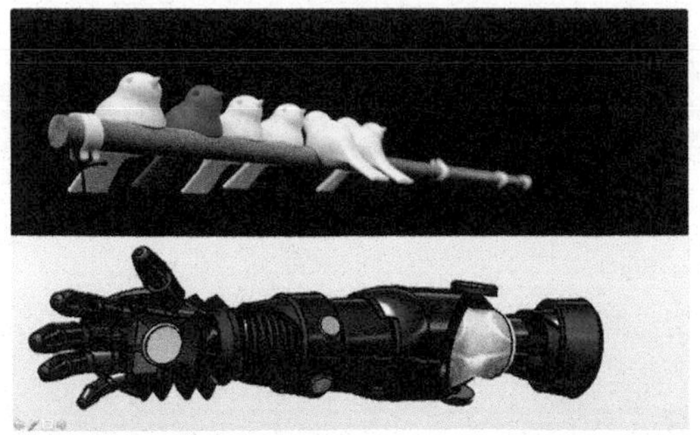

Figure 4-4 *More examples of models created using 123D Design*

Download and Set Up

If you have not yet done so, download 123D Design from 123Dapp.com (Figure 4-5) and install it on your computer. As this book goes to print, the available versions are PC, Mac, and iPad.

Figure 4-5 *The 123D Design download page on 123Dapp.com*

The first time you open the program, you'll be presented with a welcome screen—actually, the beginning of a slideshow of Quick Start Tips (Figure 4-6). Take a minute to look over these nine slides to get familiar with 123D Design.

When you're done, you can sign in to 123D Design using your existing 123D account, create a new one, or simply click the big button to

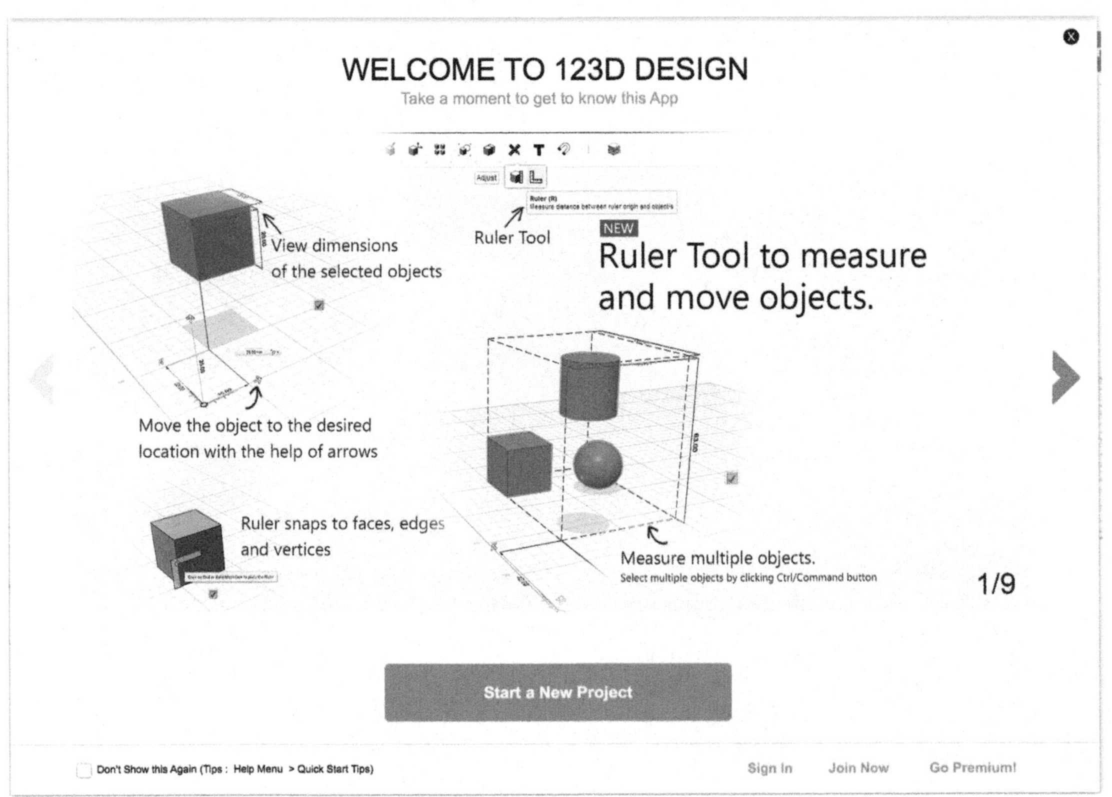

Figure 4-6 *The 123D Design welcome screen*

Start a New Project (in which case, you can always sign in to Design later on).

Getting Around

While we are still playing around and getting to know the app, let's practice using the three-button mouse to move around the space.

Roll the middle scroll wheel to zoom in and out (Figure 4-7).

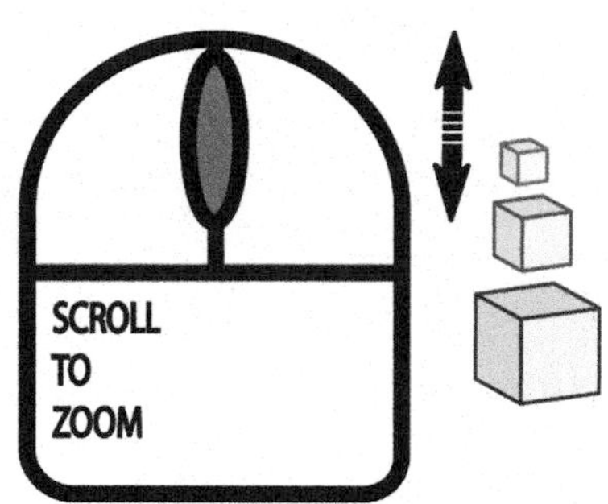

Figure 4-7 *Roll the middle scroll wheel to zoom to and away from the object*

Hold the scroll wheel down and move the mouse to pan (Figure 4-8).

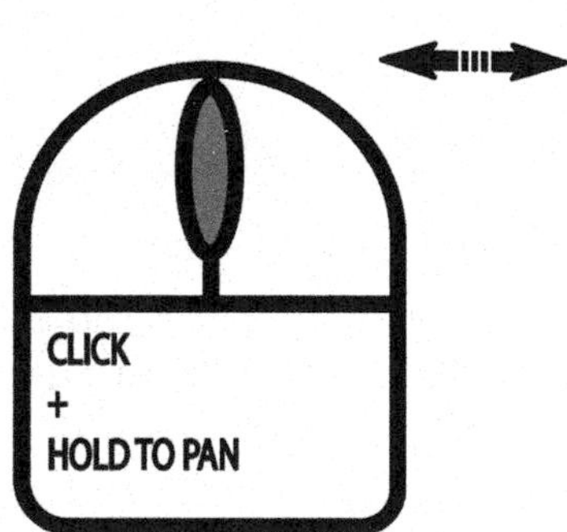

Figure 4-8 *Panning moves your view horizontally along the object*

Right click and hold to orbit (Figure 4-9).

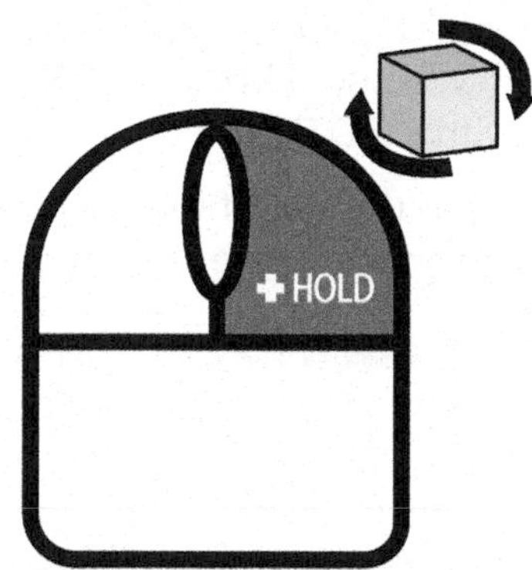

Figure 4-9 *Right click and hold down to circle around the object*

Getting Started: Playing with Blocks

Creating in this program is essentially a simple two-step process repeated over and over again (Figure 4-10):

1. Make a sketch, or create a two-dimensional shape.
2. Apply an action or feature such as "extrude" or "revolve" to turn that sketch into a three-dimensional shape.

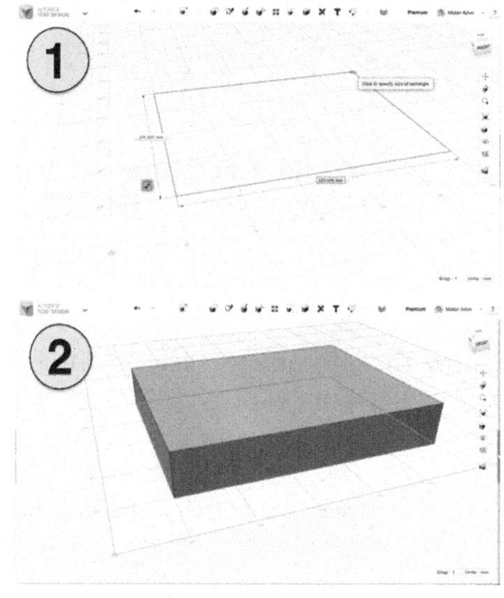

Figure 4-10 *Making a model in 123D Design is essentially a two-step process*

That's it! With this knowledge alone (plus a bit of practice) you should be able to build almost anything with 123D Design.

Before reading the rest of the chapter, open up 123D Design and let's give this a try.

Step 1: Create a Rectangle on the Main Plane

1. Hover your arrow over the top menu bar until you see the yellow Sketch prompt appear (Figure 4-11).

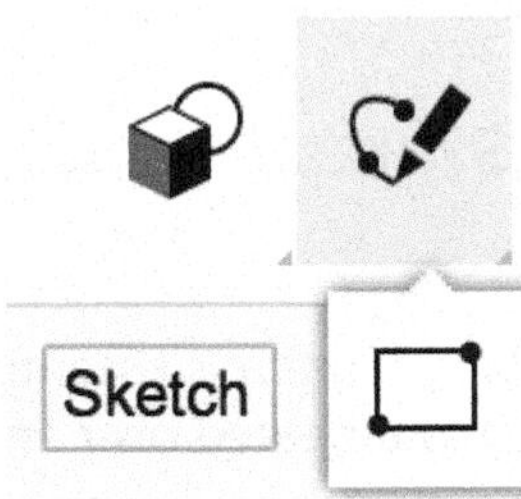

Figure 4-11 *Sketch prompt*

2. Choose Sketch Rectangle (Figure 4-12).

Figure 4-12 *Choose the Sketch Rectangle option*

3. Follow the prompt to click on the big blue grid (Figure 4-13).

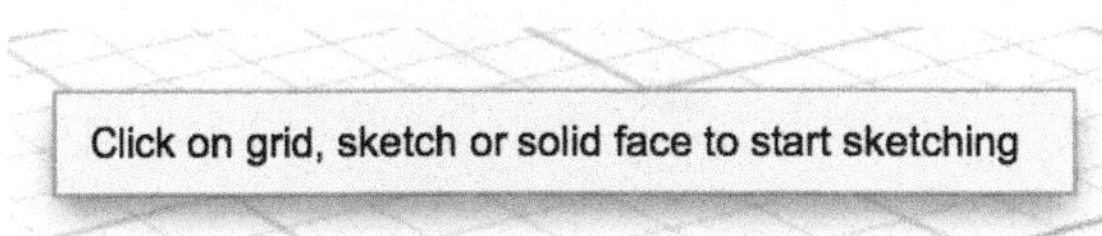

Figure 4-13 *Click on blue grid*

4. Click once to start the rectangle (Figure 4-14).

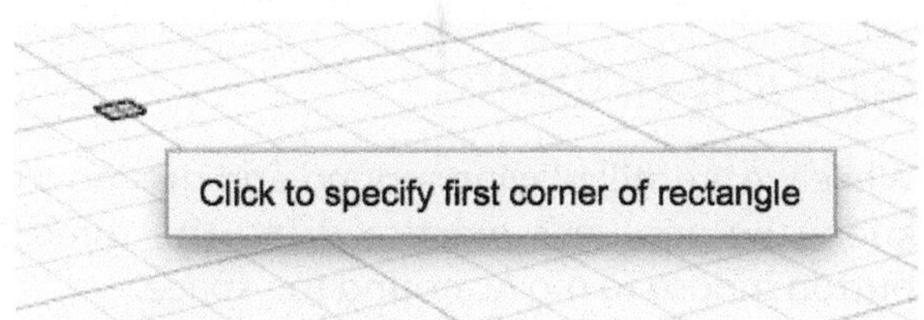

Figure 4-14 *Start the rectangle*

5. Move your cursor and click again to stop (Figure 4-15).

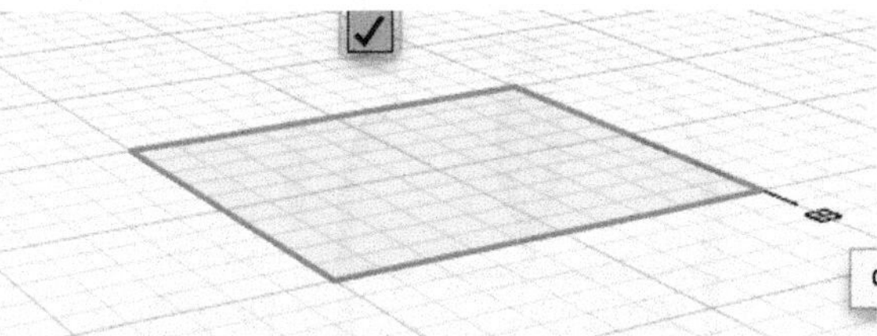

Figure 4-15 *Click to stop*

6. Select the green checkbox to exit (Figure 4-16).

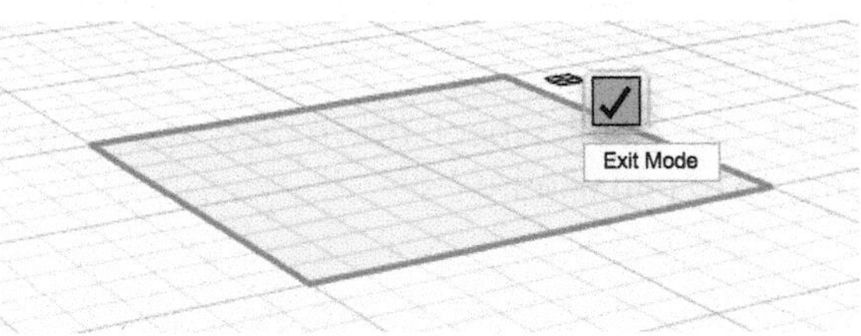

Figure 4-16 *Click the green box to exit*

Step 2: Extrude the Rectangle into a 3D Object

1. Hover the cursor over the top menu bar and select Construct (Figure 4-17).

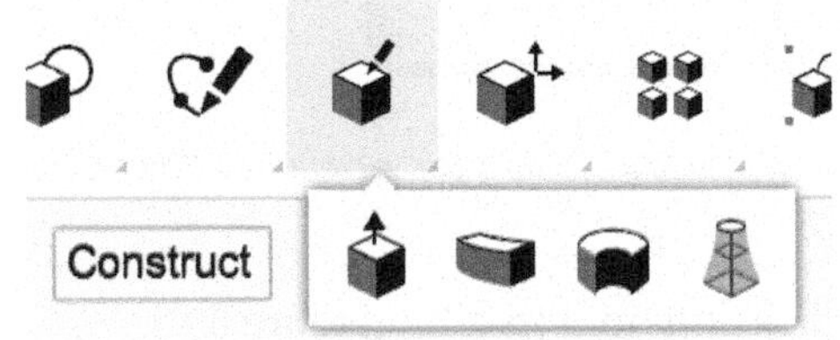

Figure 4-17 *Select the Construct option*

2. Then select Extrude (Figure 4-18).

Figure 4-18 *Select the Extrude option*

3. Follow the prompts to select your 2D rectangle (Figure 4-19), and then use the arrow by clicking and holding while moving your cursor (Figure 4-20).

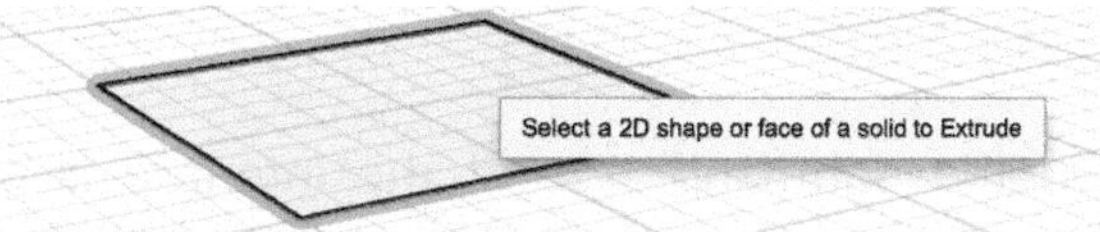

Figure 4-19 *Select your rectangle*

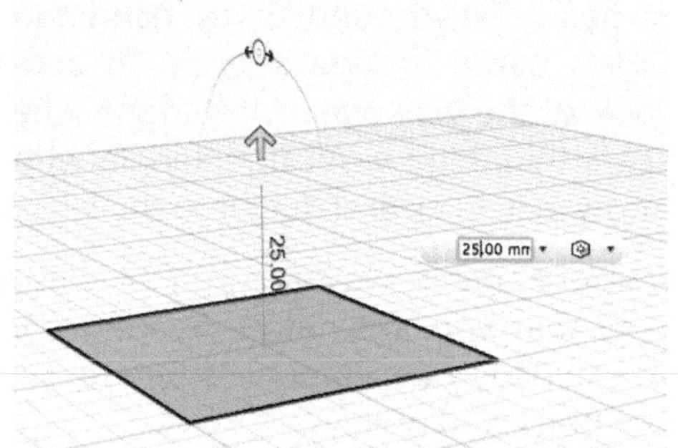

Figure 4-20 *Click and hold*

4. Release, and you now have a box (Figure 4-21).

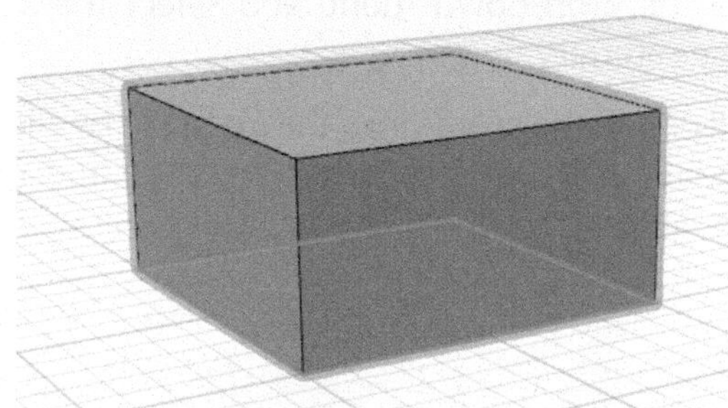

Figure 4-21 *Your new box!*

Congratulations—you are now a CAD modeler!

Take a few minutes right now to practice these two steps a few more times, using various sketch shapes and construction commands.

Now, let's go through the details so we can become experts.

Step 3: Make Changes to the Model

You have a box. How do you select the elements of the box that you would like to alter?

1. To select an object, simply click on it (Figure 4-22).

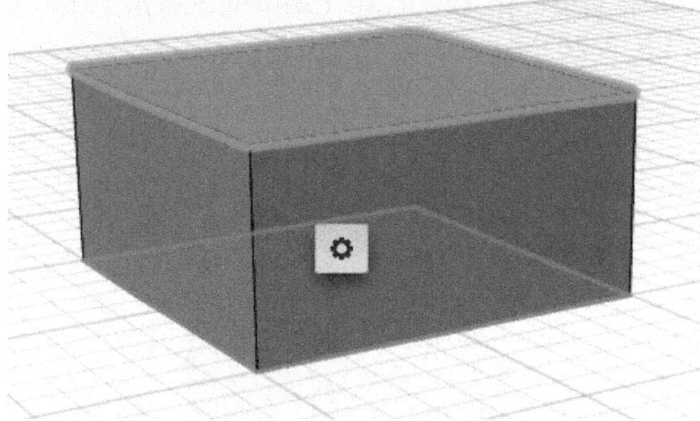

Figure 4-22 *Click to select*

2. To select a specific face, edge, or corner, click again on the element you wish to select (Figure 4-23).

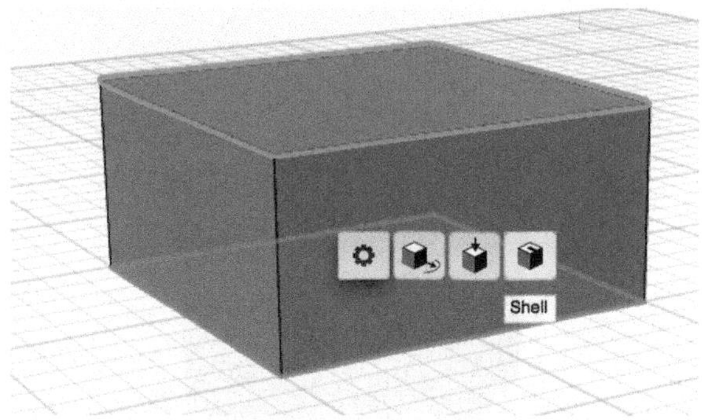

Figure 4-23 *Click the element*

3. To alter, hover over the gear icon and select a method to alter your selection (Figure 4-24).

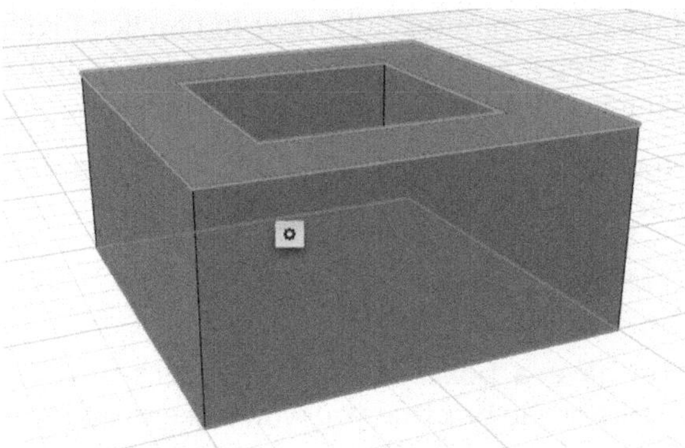

Figure 4-24 *Select a method to alter*

Spend a few minutes playing around here, to see if you can replicate the images in Figures 4-25, 4-26, and 4-27.

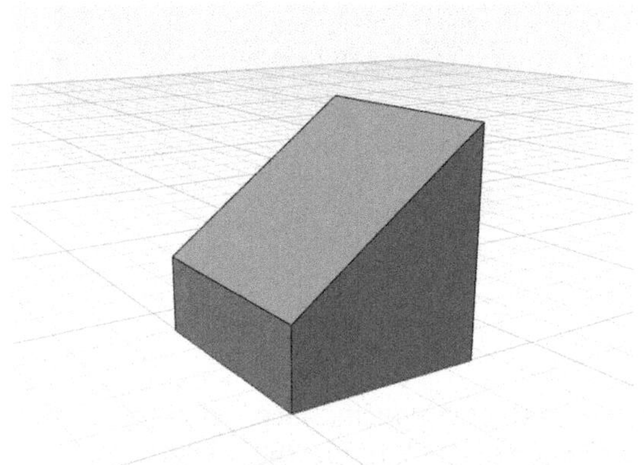

Figure 4-25 *Test Object A*

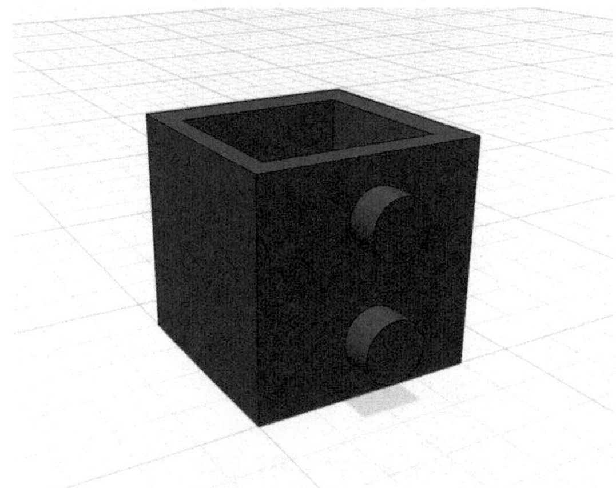

Figure 4-26 *Test Object B*

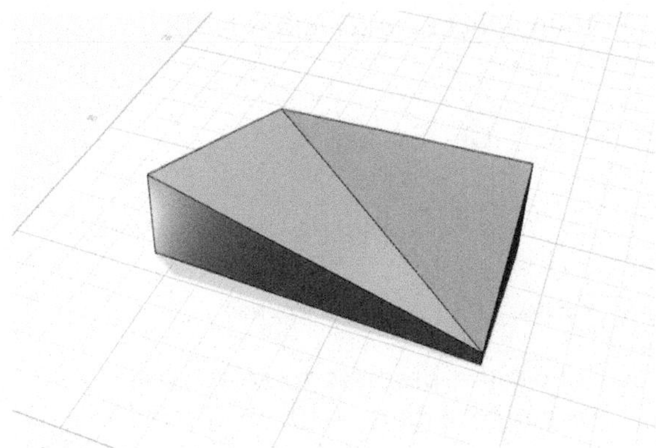

Figure 4-27 *Test Object C*

The Parts Library

On the righthand side of the interface is a library full of premade parts that you can utilize for your creations. This library, which is regularly updated, includes things such as lighting fixtures, gears, playground parts, hand tools, and models based around science. To access them, look all the way over to the right, where you'll see a thin, blue vertical rectangle with a white arrow in it. Click and drag it to the left.

This is also a great way for you to design quickly so that you are not re-creating parts that have already been modeled before, such as screws, bolts, and pulleys.

Let's quickly build a model of a molecule—a collection of atoms and their chemical bonds, represented here by balls and sticks—before we begin modeling from scratch.

Use the drop-down menu and select the Science category.

Pull one of the atoms into the interface by clicking on it and dragging it over near the grid. Do the same with a bond (Figure 4-28).

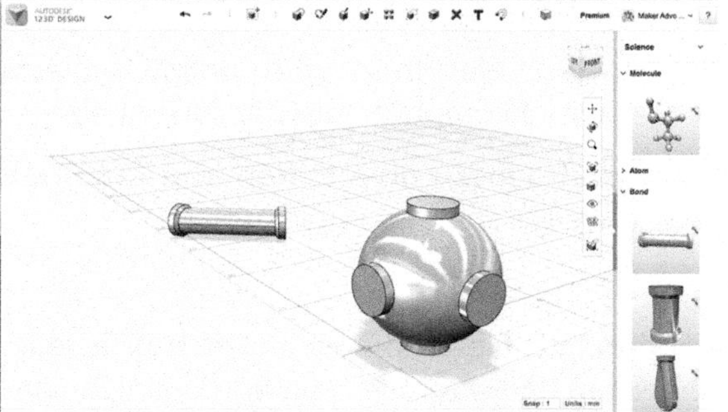

Figure 4-28 *Place an atom and bond on the interface*

Choose the snap tool from the end of the toolbar menu. Follow the prompts to select an end circle of the bond, and then an end circle of an atom (Figure 4-29 and Figure 4-30, respectively).

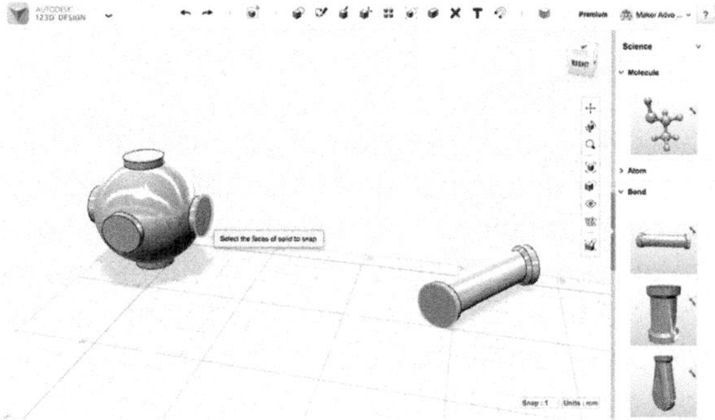

Figure 4-29 *Select a bond and an atom*

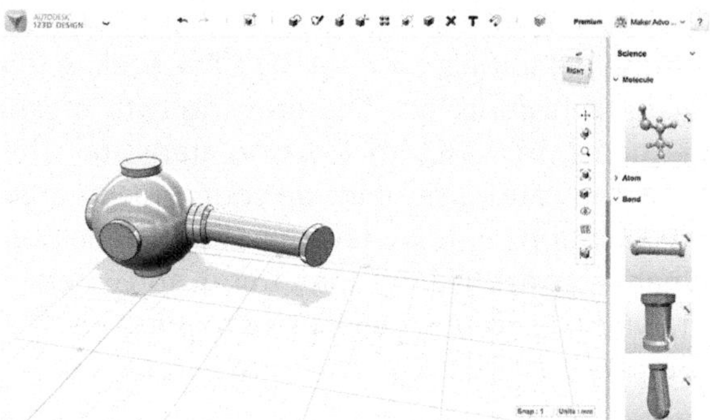

Figure 4-30 *Attach!*

Continue adding bonds and atoms until you form a molecule (Figure 4-31).

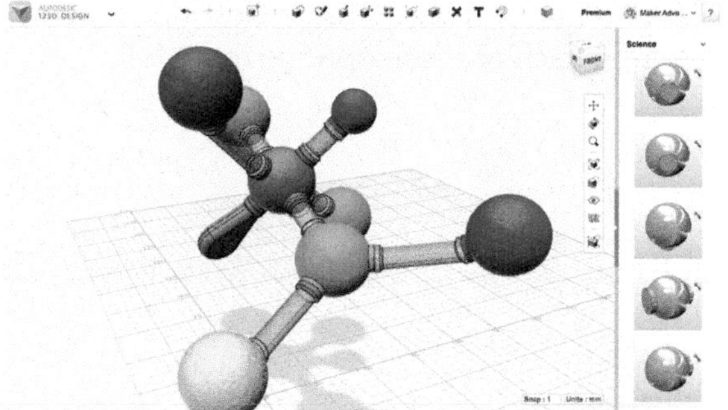

Figure 4-31 *A molecule (or futuristic space station)*

If you select the molecule, go to grouping and then select Ungroup All. You can then add different materials to each piece of the structure.

Try this with other parts in the library. It's a fun, easy way to learn how to navigate the program.

Now, let's get into the thick of it!

Start Building from Scratch

When you begin a new project or reopen an existing project, you'll be working on the *main plane* (Figure 4-32). This big blue grid is the surface or workspace where you place objects and sketches, and manipulate them to create and fine-tune your design.

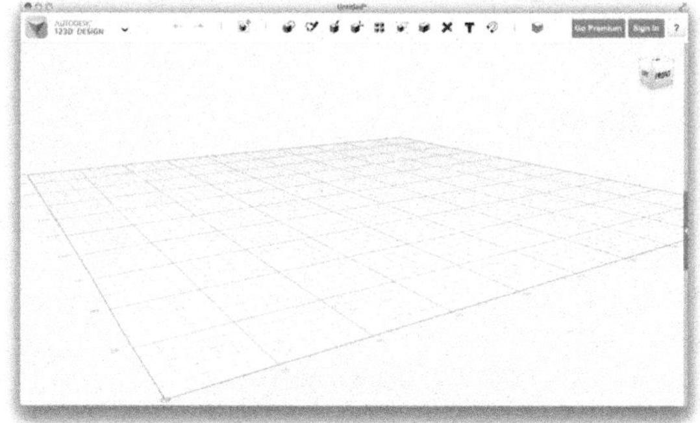

Figure 4-32 *The main plane in 123D Design*

You can think of the main plane simply as a piece of stiff paper that you can draw upon.

Let's get familiar with the menu options across the top of the 123D workspace, above the main plane (Figure 4-33). The following sections describe each of these options more fully.

Main Menu

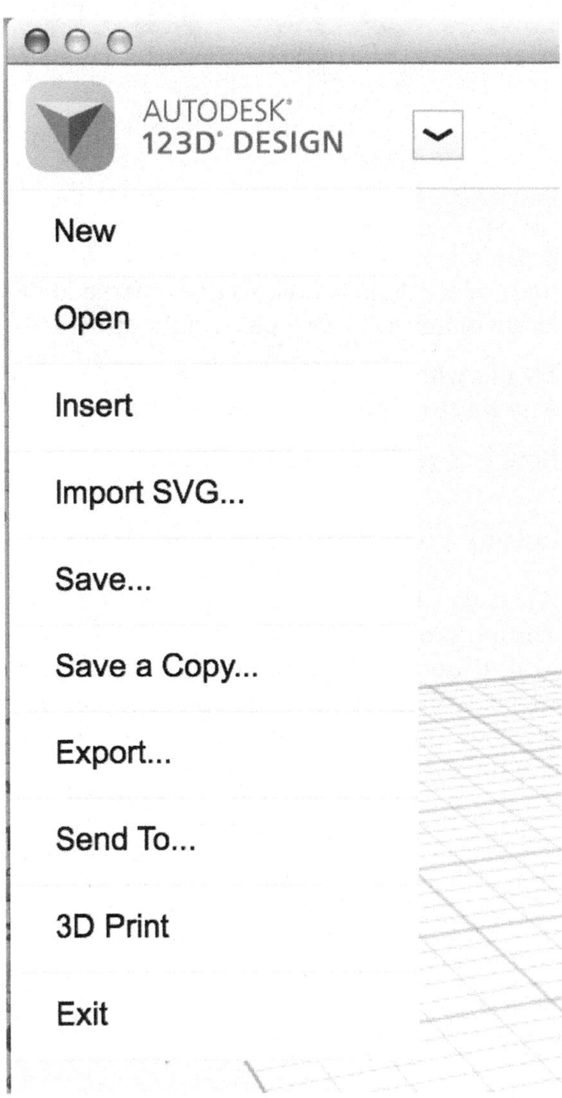

Figure 4-33 *The 123D Design main menu*

New

Starts a new file.

Open

Opens a preexisting file. You can also use this option to choose one of the thousands of models in the 123D online gallery.

Insert

Imports STL files from your online projects space or your computer into the current file.

You can also insert objects from the 123D Gallery, as well as some example files.

Import SVG...

Imports SVG files either as a sketch or a solid. These can be vector files you find online or create in other programs such as Adobe Illustrator, Corel Draw, or Inkscape, or higher-end design programs such as Autodesk Inventor, SolidWorks, or Rhino. Once imported, you can manipulate these files in 123D Design.

Save...

Saves your sketch, either in the cloud or to your computer.

Save a Copy...

Saves the current version of your design while keeping the original intact. Use this as a way to do rapid iterations. You have two options: Save "to my projects," which will save your project to your 123D account in the cloud; and "to my computer," which saves your project offline, to your own hard drive.

Export...

Save to STL, SAT, or create a 2D layout. STL files are the default file type for 3D CAD work. A 2D layout is historically a line drawing with measurements. In 123D, this option translates the file for you and saves it inside your projects area on 123Dapp.com. 2D layouts are very helpful when you're building something and need a drawing file. (It may take a few minutes for the drawing to appear, so be patient.)

Send To...

Options include:

- 123D Make, where you can slice it up into flat pieces for fabrication.
- CNC utility, which you can use to cut your file subtractively with a ShopBot or other CNC router device.
- 3D print service, which opens your file on 123Dapp.com so you can choose between various services that can print your object into reality: 3D Hubs, i.materialise, Sculpteo, and Shapeways.

3D Print

Opens your file in 123D Meshmixer where you can check, heal, and support your file, and set it up for at-home 3D printing. You can also save it as an STL file and send it out for a 3D print after you've run it through a healing operation.

Exit

Closes 123D Design.

The Ribbon Menu

The ribbon, or control panel (Figure 4-34), contains tools that allow you to create and alter shapes and volumes located on the main plane.

Figure 4-34 *The ribbon, or control panel*

Undo/Redo

The left- and right-facing curved arrows (Figure 4-35) allow you to Undo and Redo actions, respectively.

Figure 4-35 *The Undo and Redo arrows in 123D Design*

Transform

The icon of a cube inside a dotted-line square (Figure 4-36) contains Transform tools you can use to move, rotate, and scale your object up or down. But you need to have an object in the workspace to transform! So let's skip these tools for the moment.

Figure 4-36 *Transform tools*

Action Toolkits

The next several icons along the top menu hold selections of tools that allow you to create, manipulate, and refine your 3D design. As you roll over each one, its name and the different tool options will pop up. The first few are the most essential for getting to work.

Primitives

Primitives are common shapes, such as a box or torus, that you can drop on the *plane* (Figure 4-37). Primitives save you from having to create these shapes from scratch over and over again as you apply them to an object. Click the shape you wish to use, use your mouse to move it to the plane, and then click your mouse to place it down. Also note that you can enter specific values for height, width, radius, and so on.

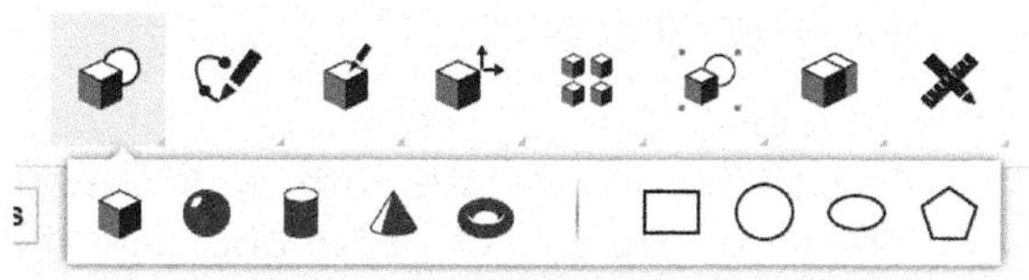

Figure 4-37 *The primitives menu*

Right-click with your mouse to move the main plane around and see different views of the objects. Zoom in and out by turning the mouse's scroll wheel. Pan (move up and down) the main plane by clicking and holding the scroll wheel while moving the mouse.

Sketch

Creating a sketch in software is similar to creating one using pen and paper. In both cases, you need something to draw *on* and something to draw *with*.

In 123D Design, the something to draw on is the main plane, as well as any flat surface on the main plane, such as the side of a cube.

The things to draw with are the Sketch tools. These come in various shapes and sizes and can be effective in different ways, depending on the complexity of the object you're making. You can choose to create common shapes such as rectangles, circles, lines, or splines, and enter specific values for their dimensions.

When you're making a sketch, be sure to include two things:

Measurements
: Even if you can only make a best guess at the final dimensions, it's a good idea to include measurements in your sketch file. Try to keep them whole numbers or one place past the decimal. More complicated numbers will be more challenging to alter later, or to match up with if you need to make another object fit into the one you're working on.

Closed shapes
: Any shape you create with your sketch has to be devoid of holes and gaps in order for you to turn it into a 3D solid.

To start sketching, select the Sketch toolkit from the action tools. Here you'll find common tools such as line, rectangle, circle, and spline (Figure 4-38).

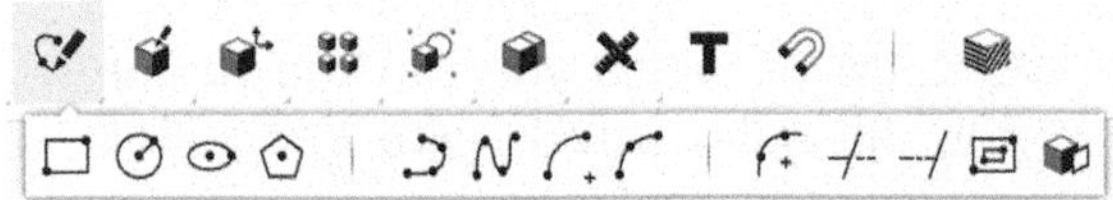

Figure 4-38 *The Sketch menu*

When sketching, keep your view cube turned so that you're looking at the sketch straight on. If you don't, you'll have a skewed perspective that will make it harder to get the shape you want as you work on the design.

Rectangle

The Rectangle is one of the tools you'll use the most as you're designing. To use it:

1. Select the Rectangle tool from the Sketch toolkit (Figure 4-39).

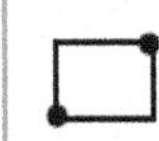

Figure 4-39 *The Rectangle tool in the Sketch toolkit*

2. Select your sketch plane: the main plane or the face of an object.
3. Click once in the place where you'd like one corner of the rectangle to be. Move your mouse until the rectangle is the size you want, then click again to finish the shape (Figure 4-40).

If you want to make a rectangle of a specific size, hit the Tab key on your keypad and toggle between width and length dimensions.

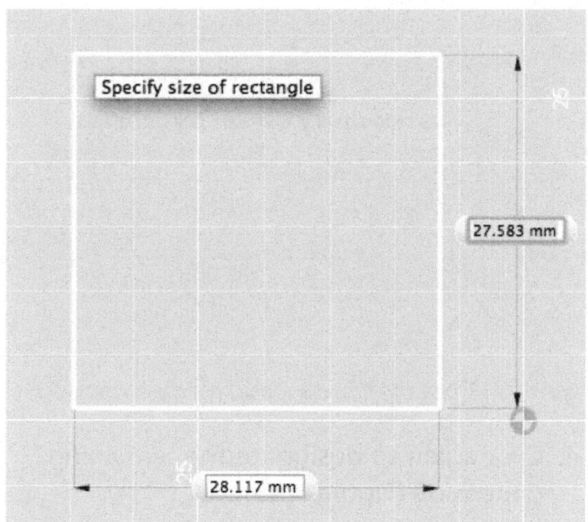

Figure 4-40 *A sketch of a rectangle on the main plane*

You can also use the Rectangle tool to generate more complex shapes by layering different rectangles on top of each other. You can then add your choice of feature to create your 3D shape.

If you need to clean up the design, use the Trim tool (Figure 4-41) in the Sketch toolkit to trim overlapping lines and create a smooth shape.

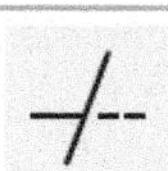

Figure 4-41 *The Trim tool*

But you don't have to: 123D Design allows you to create a complex 3D shape even if it's not perfectly trimmed.

You can extrude an overlapped shape by selecting which parts you want or don't want extruded.

You can also use the Rectangle tool to set up basic parameters for your complex shape—for example, to keep your design within a certain number of inches without going beyond that measurement.

Circle

To use the Circle tool:

1. Select the Circle tool from the Sketch toolkit (Figure 4-42).

Figure 4-42 *The Circle tool*

2. Click on the main plane (the blue grid) where you'd like the center point of the circle to be (Figure 4-43).

Figure 4-43 *Click where center of the circle should be*

3. Move your mouse away from the center point until the circle is the size you wish, and then click again to "set" that size. You can also type a specific dimension into the dialog box and then press the Enter key (Figure 4-44).

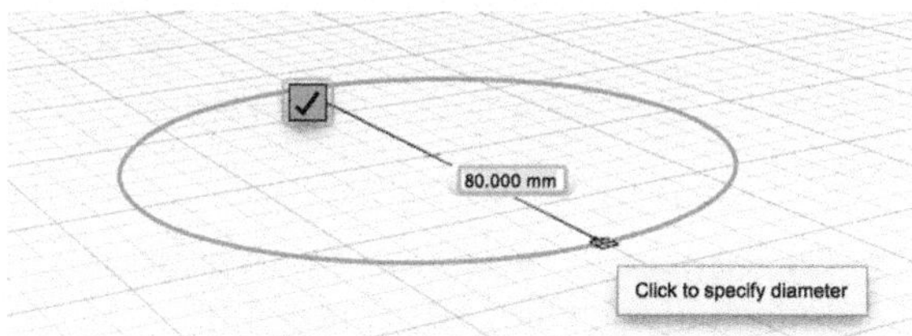

Figure 4-44 *Set the size of the circle*

4. Click the checkmark to finish sketching your circle.

You can use the Circle and Rectangle tools together to make a whole lot of different designs. For example, you can add rounded corners to a rectangle, or create a rectangular shape with a circular shape cut out of it.

Ellipse

The Ellipse tool makes it easy to add a bit of an arc to a surface. You may find yourself using this tool frequently. The Ellipse tool is also helpful when doing the loft operation (shown below).

To use the Ellipse tool:

1. Select the Ellipse tool from the Sketch toolkit (Figure 4-45).

Figure 4-45 *The Ellipse tool*

2. Click the plane where you want the center of the ellipse to be (Figure 4-46).

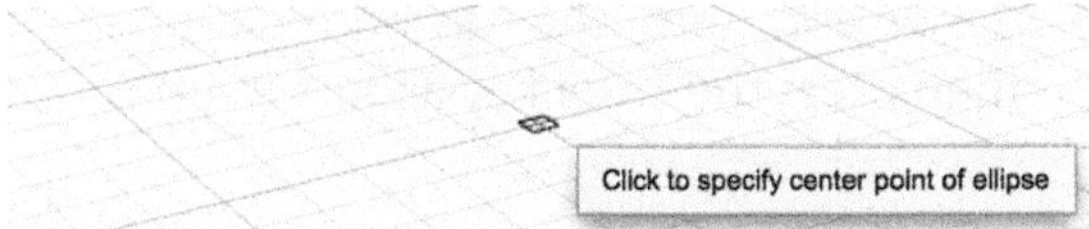

Figure 4-46 *Click where the center of the ellipse should be*

3. Click again to designate the height or length of the ellipse (Figure 4-47).

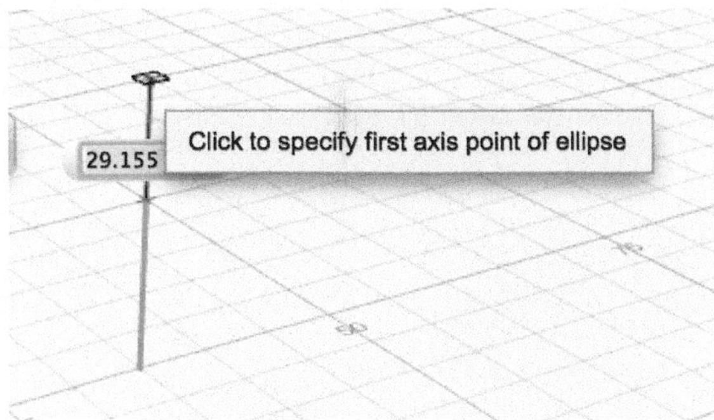

Figure 4-47 *Set the first dimension of the ellipse*

4. Click again to designate the remaining dimension (Figure 4-48).

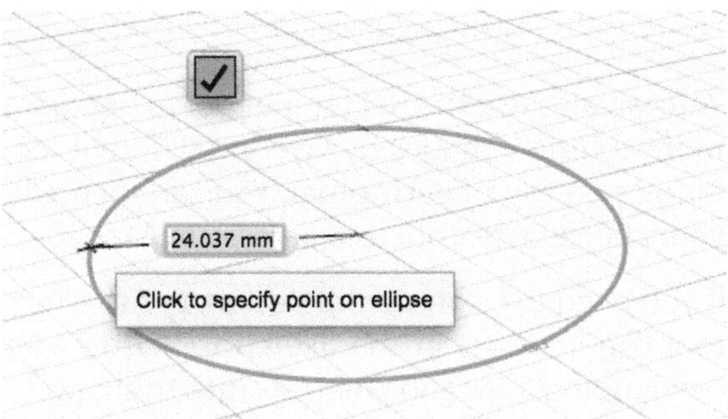

Figure 4-48 *Set the other dimension of the ellipse*

Polygon

The Polygon tool allows you to create multisided shapes.

To use the tool:

1. Select the Polygon tool from the Sketch toolkit (Figure 4-49).

Figure 4-49 *The Polygon tool*

2. Click on your sketching surface where you want the center of the polygon to be.
3. Move the mouse to the desired radius and click again.

4. To change the number of sides, press the Tab key and toggle between dialog boxes, and type in the number of sides you want. Then click again to end the sketch.

Need to make a perfect triangle? Use the Polygon tool.

Polyline

Use the Polyline tool to create line segments that make up shapes. A line segment consists of a straight or curved line that joins two points. Once you've created a shape, you can apply a feature, such as extrusion or loft, to the shape you've created. The Polyline tool, a well as all other sketching tools, allows you to add a third dimension to your sketch and make precise shapes that can be fit together.

To use the Polyline tool:

1. Choose the Polyline tool from the Sketch toolkit (Figure 4-50).

Figure 4-50 *The Polyline tool*

2. Click once where you'd like your line to start. Release the mouse, move your mouse to the location where you want your line to end, and click again.
3. To make multiple segments, move the mouse and click again.
4. To create an arc using the Polyline tool, hold down the left mouse button and drag, creating the radius of the arc.

How Do I Make a Shape?

To turn your lines into a shape, you must close the segment you've been sketching. Here's how to do it:

1. Bring your cursor back to the starting point of your sketch. A little square appears when the cursor is positioned correctly.
2. Click in the middle of the square to select the shape. It will turn a solid color, indicating that your segment is correctly closed.

Figure 4-51 *Make a shape*

3. Click back to your first point.

OK, now it's time to use the skills you've just learned to build something fairly simple.

Spline

Use the Spline tool when you want to make a freeform organic shape to extrude or revolve.

Think of splines this way: imagine that you have a board with nails in it and a string that you're going to thread around the nails to make curves. Each click you make in the sketching space is like putting a nail in the board, and the software automatically curves the string around it.

You can end a spline either by taking it back to its original starting point or by right-clicking or double-clicking.

Splines allow you put fun and crazy curves into your shapes—and useful curves, such as those in the body of a guitar (see Figures 4-52, 4-53, and 4-54).

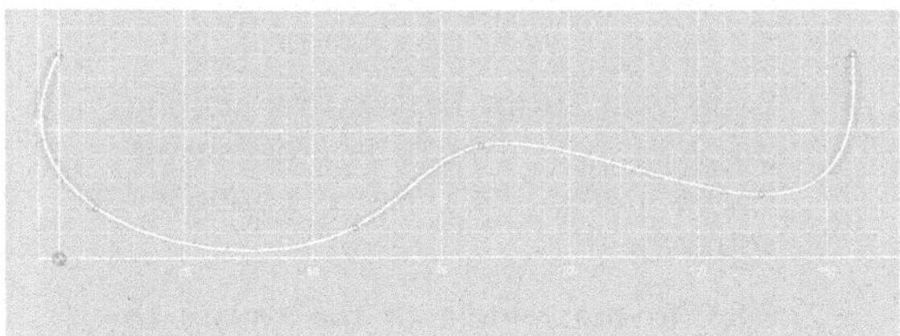

Figure 4-52 *A combination of simple splines*

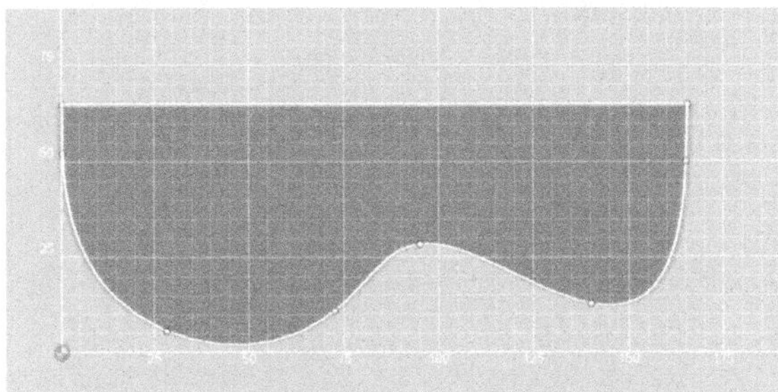

Figure 4-53 *With the segments closed, the splines become a shape*

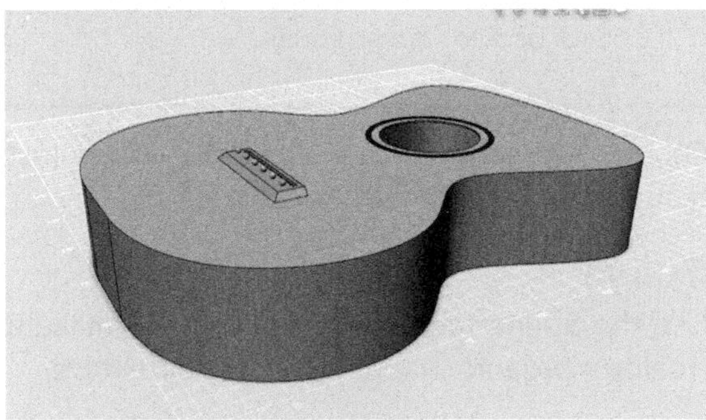

Figure 4-54 *By applying extrusion tools and other detailing, the shape becomes three dimensional*

To use the Spline tool:

1. Select the Spline tool from the Sketch toolkit (Figure 4-55).

Figure 4-55 *Select the Spline tool*

2. Click in the main plane to create your first point (Figure 4-56).

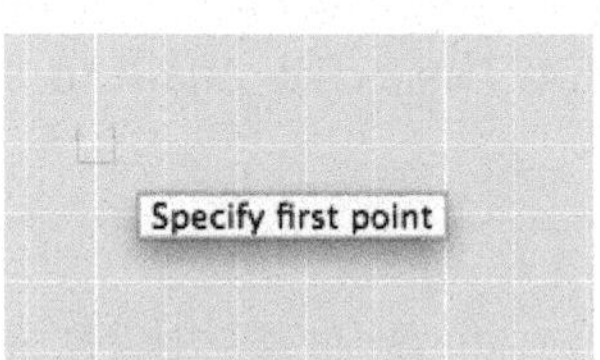

Figure 4-56 *Create the first point*

3. As shown in Figure 4-57, move your mouse and click again to create your second point. (This will show up as a straight segment—that's OK.)

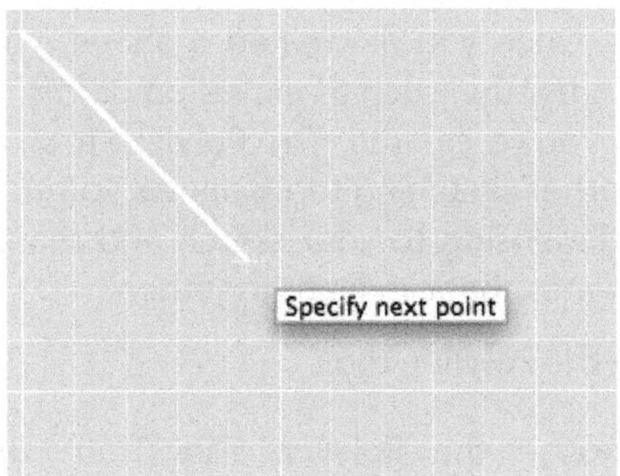

Figure 4-57 *Create the second point*

4. Move your mouse and click once more. You will now see the curve (Figure 4-58).

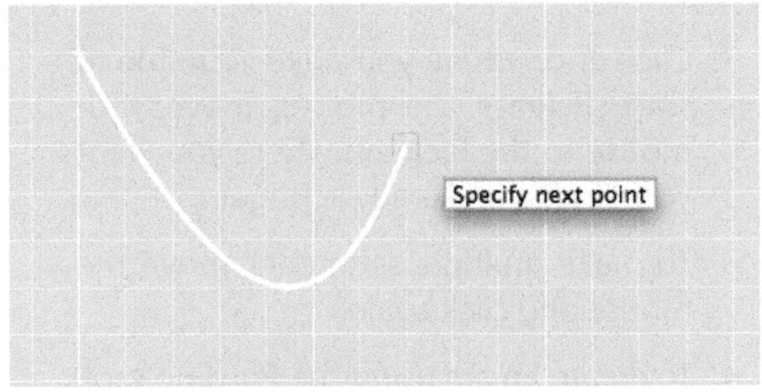

Figure 4-58 *Move the mouse and click to see the curve*

5. Continue clicking to make your desired shape (Figure 4-59).

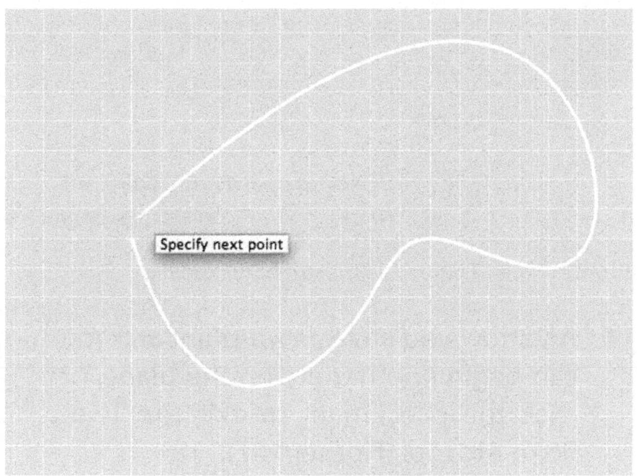

Figure 4-59 *Another click expands the shape*

The closer together your anchor points are, the tighter the curve of the spline will be.

6. If you're making an open shape, you can either double-click to end the sketch and then hit Escape, or click the checkbox to exit your sketch.
7. If you're making a closed shape, you'll see that a blue box appears when you get back to your starting point (Figure 4-60).

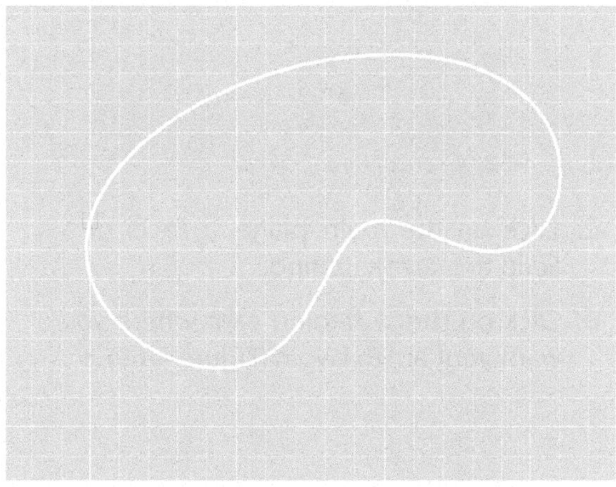

Figure 4-60 *Start the closed shape with the first click*

8. When you've created your spline, you can still tweak it by clicking and dragging on anchor points until you've got your desired shape.

Do not attempt to trace a complex shape by putting down thousands of spline points. You'll likely crash the program and then regret that you wasted so much time. Instead, build your shape using a combination of primitives and splines.

Like rectangles and circles, splines are most useful in conjunction with other shapes and geometry features (Figure 4-61 and Figure 4-62).

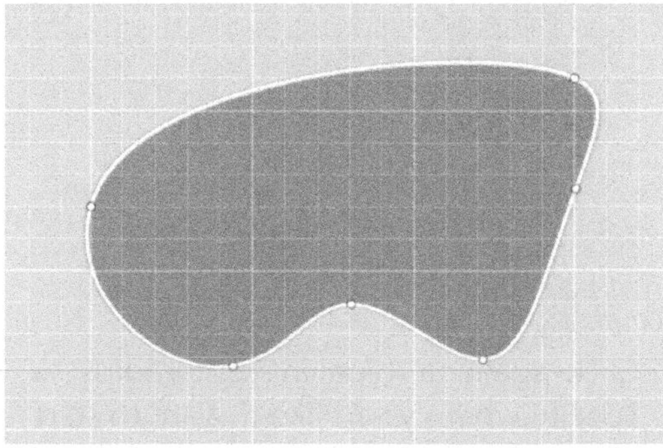

Figure 4-61 *Which primitives have been combined with splines to create this shape?*

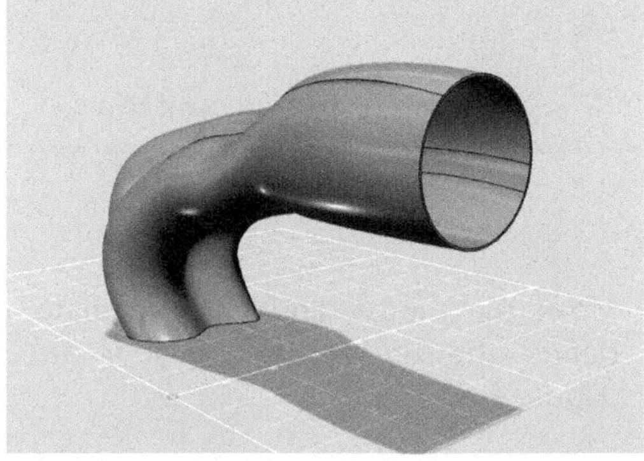

Figure 4-62 *Here, simple circles, cylinders, and splines add up to a complex 3D object*

Two-Point Arc

The Two-Point Arc tool allows you to define a curve by setting two points and adjusting the amount of curvature between them.

1. Select the Two-Point Arc tool from the Sketch toolkit (Figure 4-63).

Figure 4-63 *An arc of two points*

2. Click on your grid or solid surface to begin sketching.

3. Click once where you'd like the center of your arc to be (Figure 4-64).

Figure 4-64 *The center of the arc of two points*

4. Click again or type in your radius to specify where you'd like the arc to start (Figure 4-65).

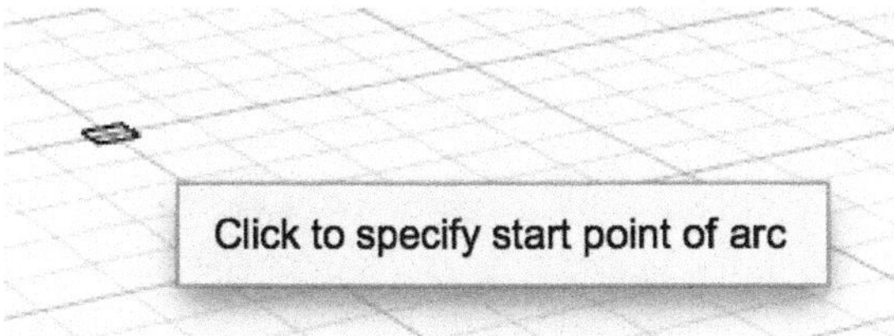

Figure 4-65 *The radius of the arc of two points*

5. Move the mouse and click again to define the end point of your arc (Figure 4-66).

Figure 4-66 *The end of the arc of two points*

6. And now you have a two-point arc! You can begin another arc on the plane, or click the checkmark to exit the Two-Point Arc tool (Figure 4-67).

Figure 4-67 *Exiting the arc of two points*

Three-Point Arc

The Three-Point Arc tool allows you to define a curve by setting three points (two end points and a middle). The software then automatically sets the amount of curvature between them.

1. Select the Three-Point Arc tool from the Sketch toolkit (Figure 4-68).

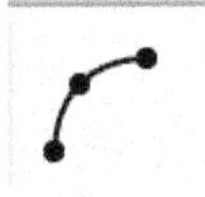

Figure 4-68 *An arc of three points*

2. Click on the main plane, a Face, or a Solid to start sketching.

3. Click on the sketching area where you want your arc to begin (Figure 4-69).

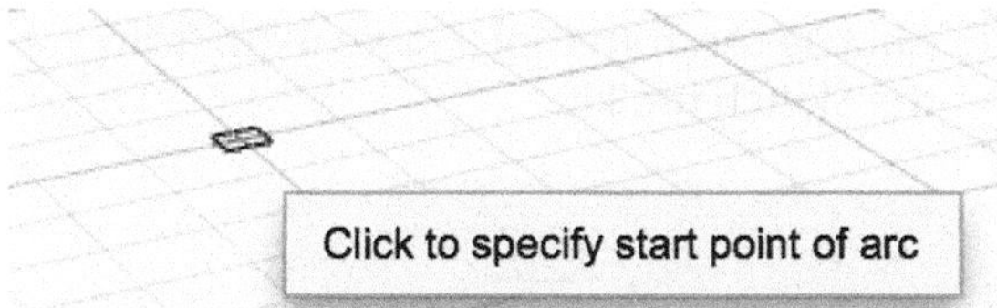

Figure 4-69 *The start of the arc of three points*

4. Click again where you want your arc to end (Figure 4-70).

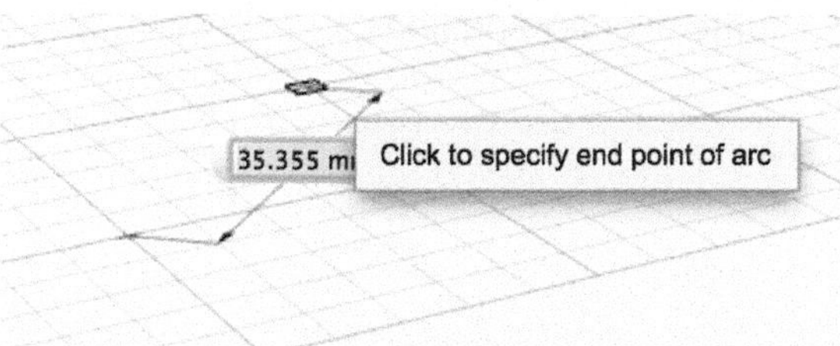

Figure 4-70 *The end of the arc of three points*

5. Click to select the third point (Figure 4-71).

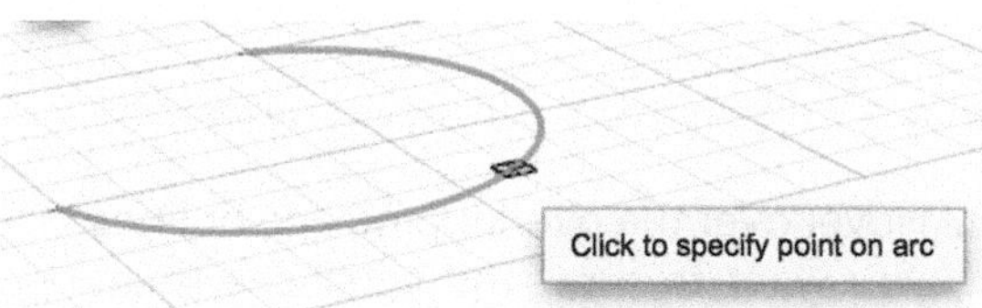

Figure 4-71 *The third point of the arc of three points*

6. You've created a three-point arc. Now you can begin a new arc, or click the checkmark to exit the three-point arc tool (Figure 4-72).

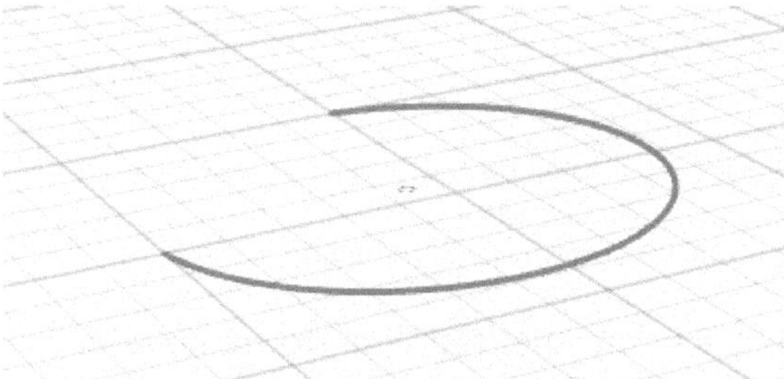

Figure 4-72 *Exiting the arc of three points*

Fillet: Adding a Rounded Corner

You can round the edges of an object with an operation called a fillet. In CAD lingo, this word is pronounced as it's written, saying the *t* at the end (experienced CAD users can always spot a newbie when they hear this term pronounced as if it's a piece of fish).

You can round edges while in Sketch or 3D Modeling mode. It depends on what order of operations you're most comfortable with, and what will work best for your model.

To fillet:

1. Choose the Fillet tool from the Sketch toolkit (Figure 4-73).

Figure 4-73 *Choose Fillet*

2. Select the sketch you want to edit, and then select one line (Figure 4-74) followed by another that you want to round (Figure 4-75). Or, select a corner that you wish to round.

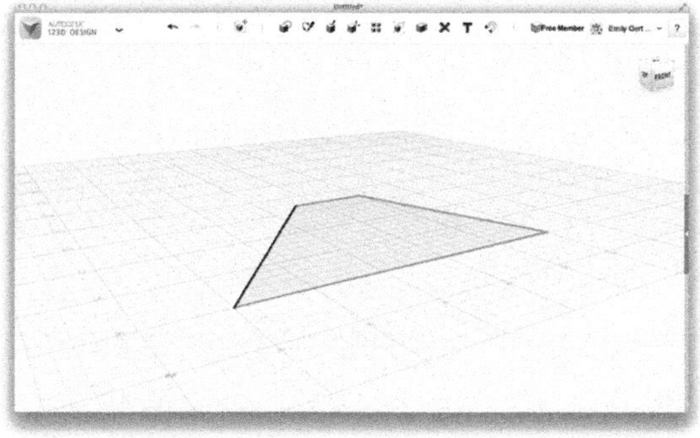

Figure 4-74 *Pick the first of two lines to round with the Fillet tool*

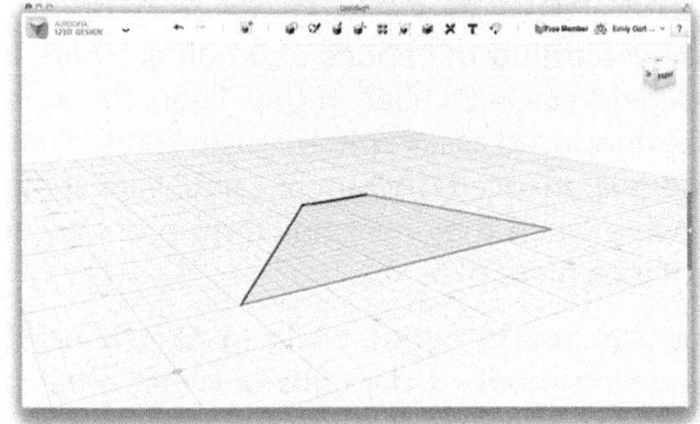

Figure 4-75 *Then pick the next line*

3. You can enter a specific radius (Figure 4-76), or push and pull the arrow (Figure 4-77) to make your fillet, using the Fillet Radius box that comes up when you select the Fillet tool.

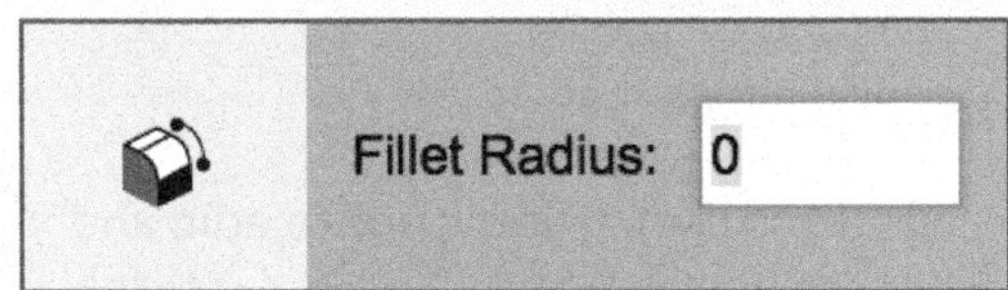

Figure 4-76 *Enter a specific radius if desired*

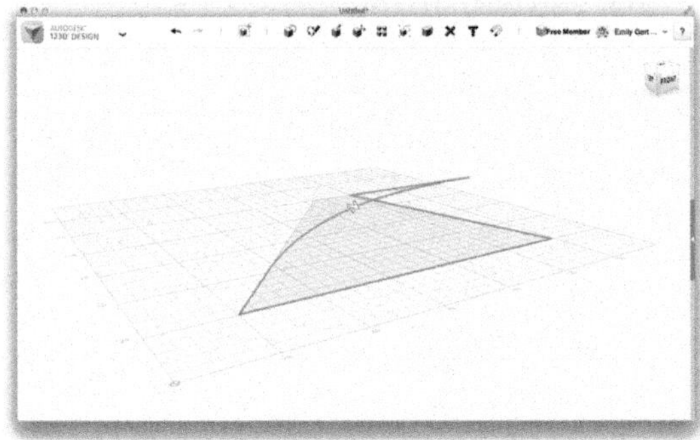

Figure 4-77 *Or, push/pull the arrow to curve the edge*

Trim

The Trim tool is very useful when creating complex shapes, or if you need an exact shape with a complex curve in it. It allows you to remove external material from your sketch. To use the Trim tool, follow these instructions:

1. Choose the Trim tool (Figure 4-78) from the Sketch toolkit.

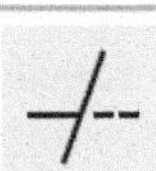

Figure 4-78 *The Trim tool*

2. Click the sketch you wish to edit (Figure 4-79). The sketch must contain some overlapping lines.

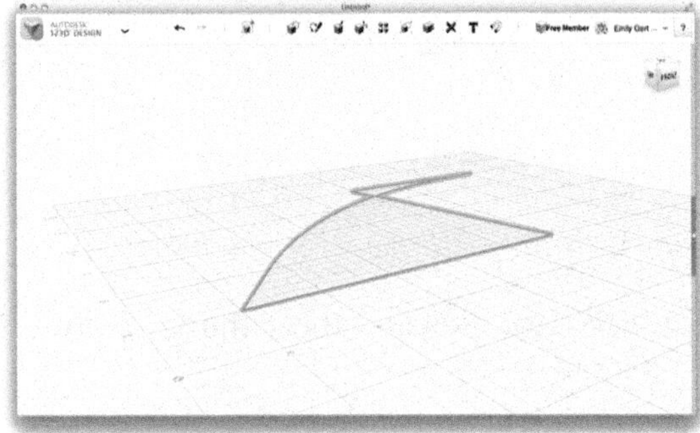

Figure 4-79 *Select the sketch you want to trim*

3. Hover over the overlapping line or lines you want to trim, and you'll see them turn red (Figure 4-80).

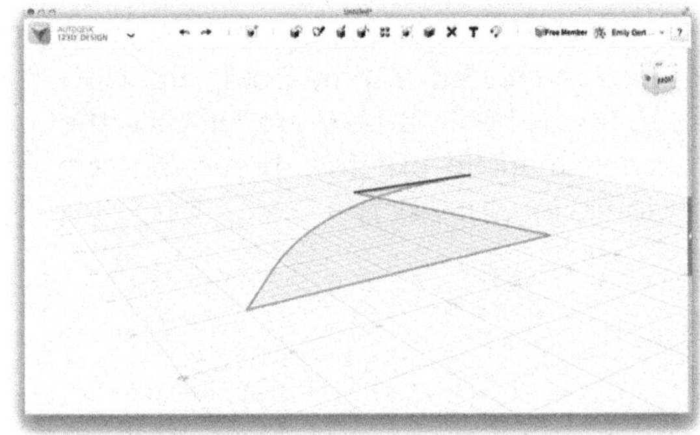

Figure 4-80 *The line to be trimmed turns red*

4. Click on the red line to remove it (Figure 4-81).

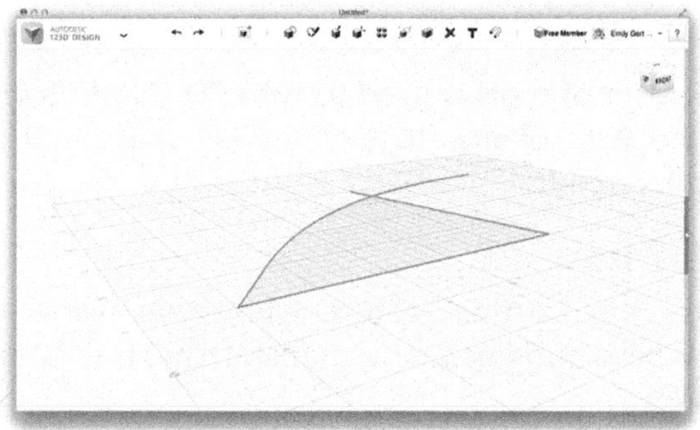

Figure 4-81 *Remove the red line with a click*

Extend

Extending a sketch is a reliable way to make sure that your sketch connects to the next intersecting line.

1. Select the Extend tool (Figure 4-82) from the Sketch toolkit.

Figure 4-82 *The Extend tool*

2. Select the sketch that you want to work with (Figure 4-83).

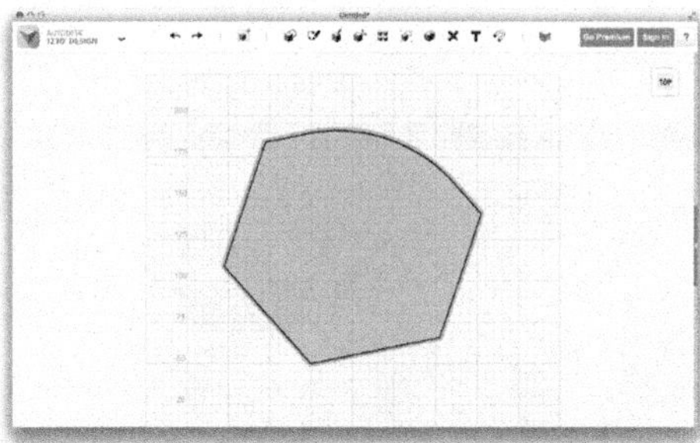

Figure 4-83 *Select the sketch that you wish to extend*

3. Select the arch within the sketch that you want to extend (Figure 4-84). A red line will appear (Figure 4-85) indicating where your sketch will extend to.

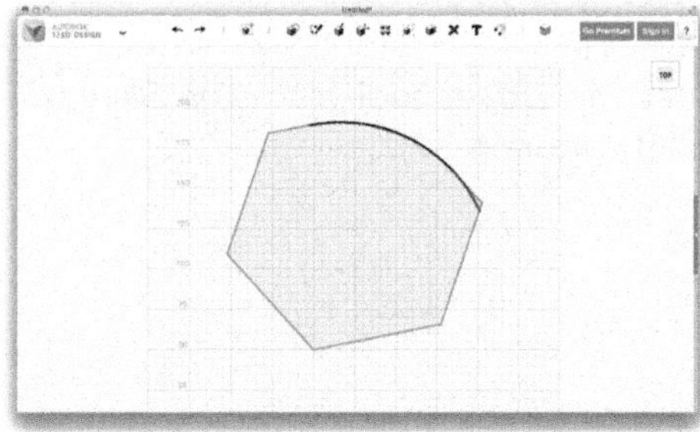

Figure 4-84 *Select the arch that you wish to extend*

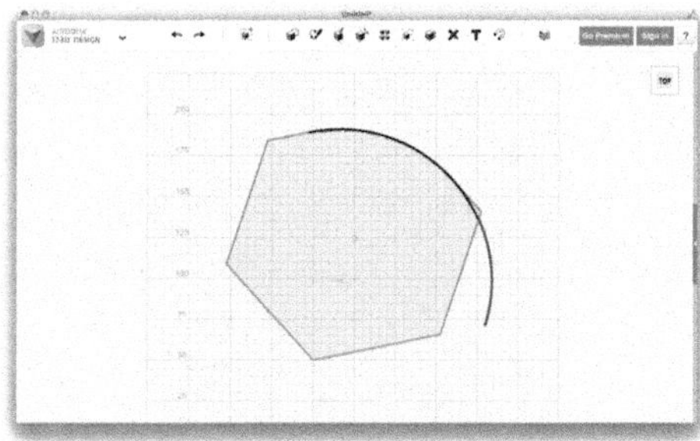

Figure 4-85 *The red line shows how far the sketch will extend*

4. Click the checkbox or hit the Enter key and the shape will become solid (Figure 4-86).

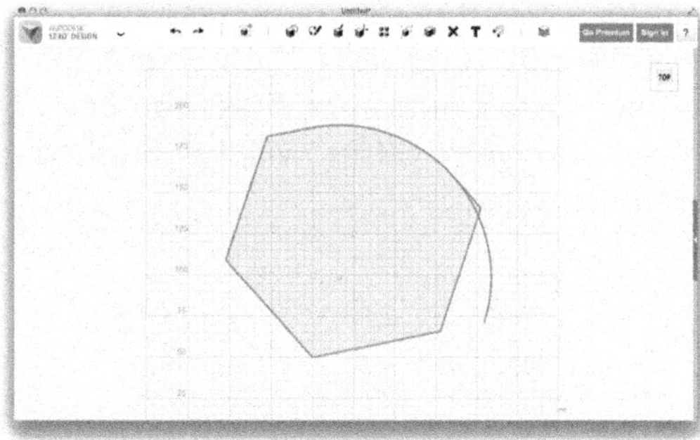

Figure 4-86 *An extended sketch.*

Offset

Offsetting lets you create thicker walls on objects with more complex shapes. In some situations, you can also do this with the Shell tool.

To use the Offset tool:

1. Select the Offset tool from the Sketch toolkit (Figure 4-87).

Figure 4-87 *The Offset tool*

2. Select the sketch that you want to offset (Figure 4-88).

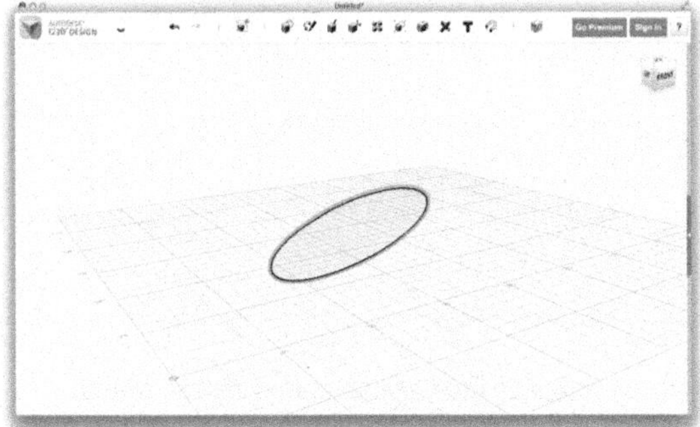

Figure 4-88 *Select the sketch to be offset*

3. Hover your mouse over a line on that sketch.
4. Enter your measurement (estimate if you're not sure), and then select the green check button or press the Enter key, and your offset will be finished (Figure 4-89).

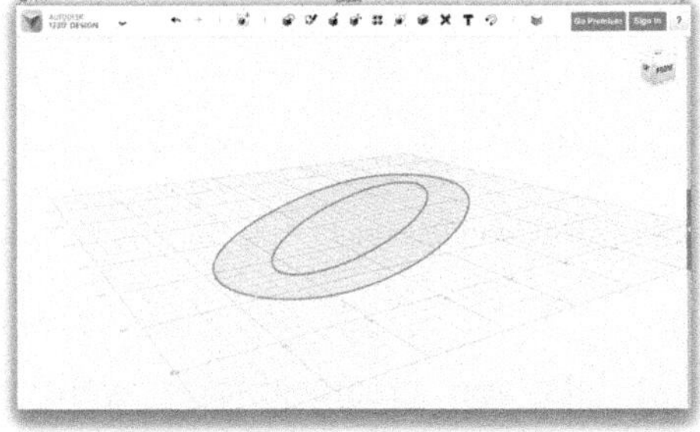

Figure 4-89 *An ellipse with an offset*

Construct

This set of tools is used to turn 2D shapes (created with Sketch tools) into 3D shapes, also called solids.

Extrude

Use the Extrude feature when you want to model a cube or shape that starts with two flat sides.

Figure 4-90 and Figure 4-91 show just a few examples of what you can do with the Extrude tool, which you'll find yourself using almost as much as the sketching tools.

Extrusion means to take a 2D sketch and add that third dimension that turns it into an object. Shapes can be extruded up, down, or outward, according to the measurements you desire.

Figure 4-90 *A flat, 2D image*

Figure 4-91 *The flat image, extruded into the third dimension*

One particularly useful aspect of the Extrude tool is that it allows you to create models from premade parts. You can make a premade part more complex by sketching on an already-extruded surface and then extruding what you've sketched. In this way, you can add new surfaces to the object.

You can also extrude in the opposite direction to create pockets or holes. This process is called cutting, because it cuts away surface area from the solid shape.

Here are the basics of using the Extrude tool:

1. Start with a sketch (Figure 4-92).

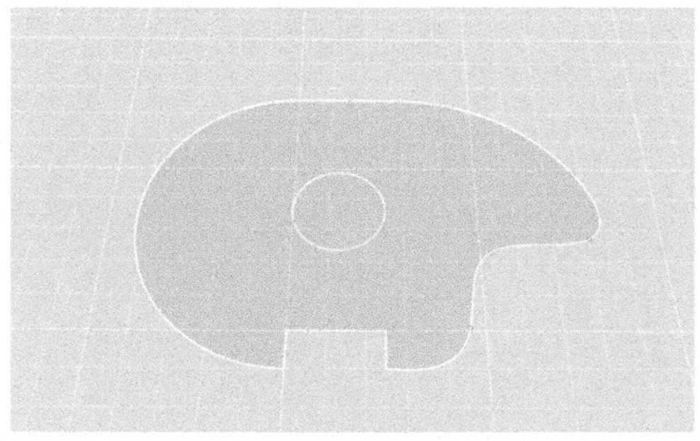

Figure 4-92 *Start the extrusion process*

2. Select the Extrude tool, shown in Figure 4-93, from the Construct toolkit.

Figure 4-93 *The Extrude tool*

3. Select the sketch, or you can select the Extrude tool from the gear symbol that appears once you've selected the sketch.
4. When you select the sketch, a little white arrow will appear, along with a dialog box. As with other tools, you can either type in specific measurements or click and drag the arrow until you have the size you want. When you're satisfied, hit the Enter key (Figure 4-94).

While you're still in the Extrude tool, you'll see a circle on an arc. You can click the circle and drag it to create a draft angle, altering the width of the top of the part. This comes in handy when making injection-molded parts. If you're planning to mass-produce a plastic injection-molded object, you'll want to use a draft angle of 2%.

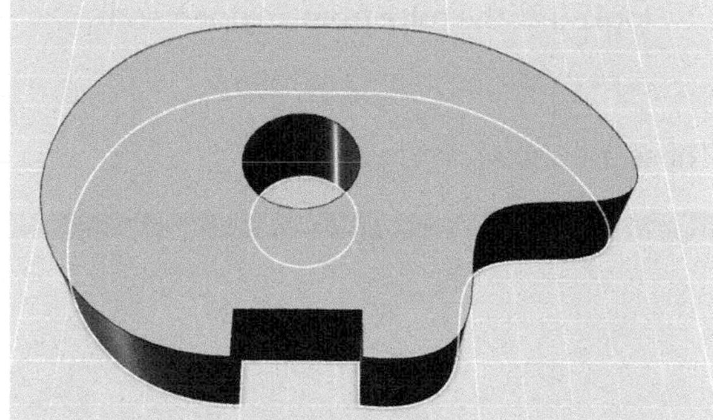

Figure 4-94 *The selected sketch*

If you want to create holes or pockets in your object, begin with a sketch on a flat surface. Select the Sketch tool and then select Flat Surface. Once you're done sketching, select the Extrude tool and pull it back through the part (Figure 4-95). You'll see the extrusion previewed in red, and you'll also see the little icon in the drop-down menu change from the join icon to the cut icon.

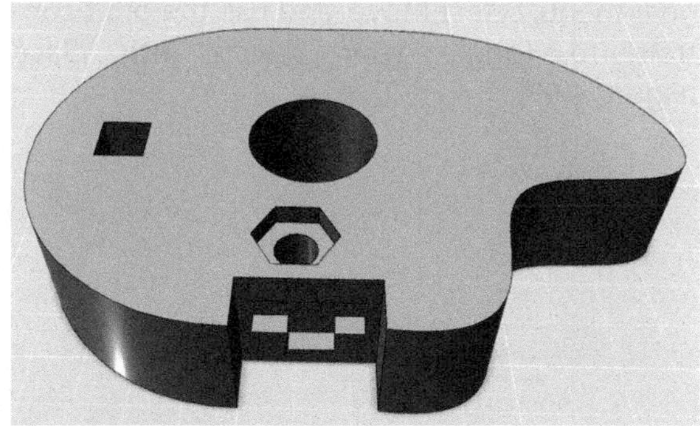

Figure 4-95 *The extruded sketch*

Sweep

Use Sweep when you want to make something like a piece of tubing or bent pipe.

Sweep is an efficient way to build more complex designs that involve tubing, such as a bicycle frame. To create a sweep, you need:

- A profile or shape for the tube (this is the shape that you would see if you looked at the tube from an open end)
- A path for the profile to follow

These are shown in Figure 4-96.

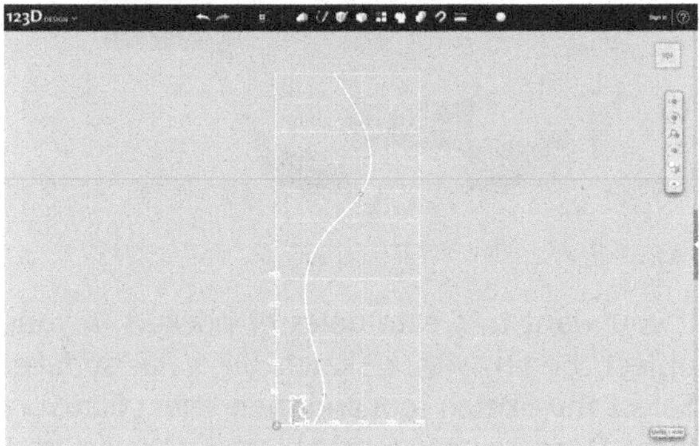

Figure 4-96 *The two sketches you will need to create for this tool on flat grid*

You'll need to move your sketch so that it's perpendicular to the sketch of your path. To do this, first select the profile sketch. In the pop-up gear menu, select Move, and use the orbit controls to toggle your sketch into place (Figure 4-97).

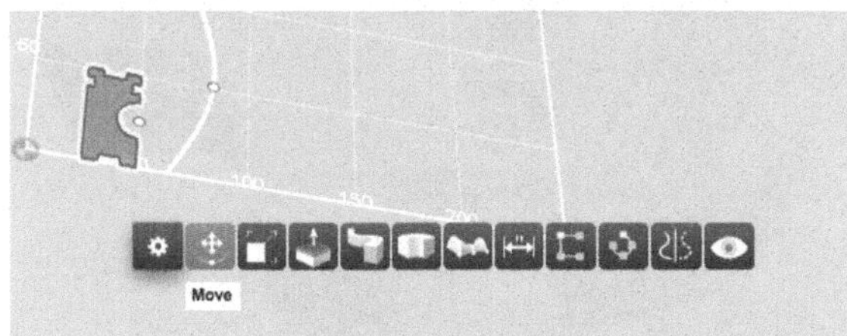

Figure 4-97 *Use the move tool to make your flat profile sketch perpendicular to your path using the Move tool*

Now, select the Sweep tool, shown in Figure 4-98, from the Construct menu.

Figure 4-98 *The Sweep tool*

A dialog box will appear, asking you to select a profile to sweep and then select a path to sweep that profile down (Figure 4-99).

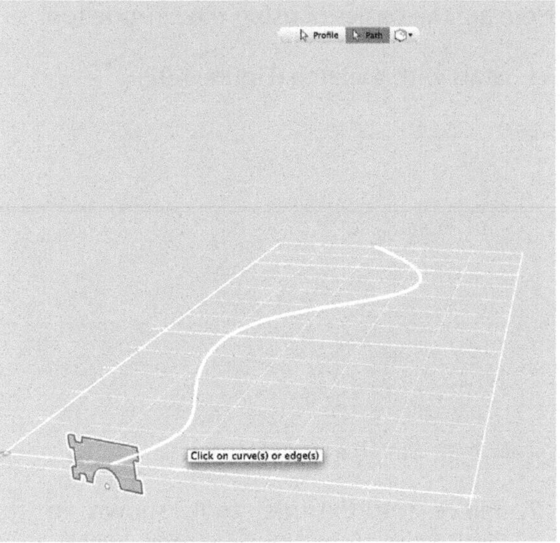

Figure 4-99 *The profile sketch perpendicular to your path sketch*

Figure 4-100 shows the result. You'll find Sweep very helpful when working on projects that have cords or bent tubing.

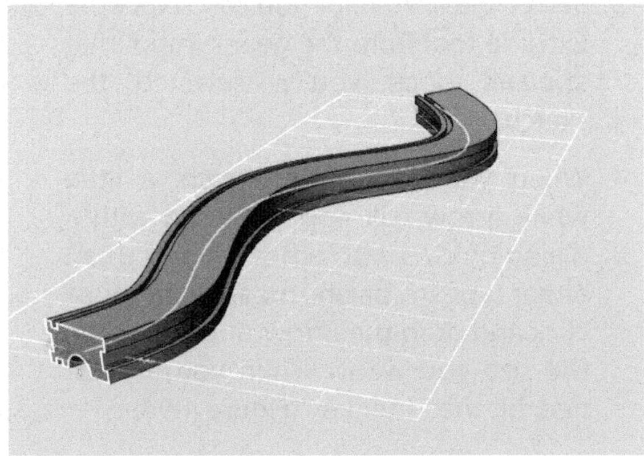

Figure 4-100 *Finished preview of the Sweep tool*

Revolve

Use Revolve when you want to make a part like a bowl, a cup, or a chess piece.

To use the Revolve tool, your sketch *must* contain a shape you want to revolve and an axis that you want it to revolve around. With simple objects like bottle or cup shapes, you can create a vertical line that will become the axis your sketch will rotate on.

This chess piece is an example of a shape that involves a closed revolve. To make it, create an outline of one-half of the final piece. Remember that in 123D Design you can have overlapping shapes and you can select multiple sketches for the same feature.

Once you've made your sketch, select the Revolve tool, shown in Figure 4-101, from the Construct toolkit.

Figure 4-101 *The Revolve tool*

When you select the Revolve feature, a little dialog box will appear. It's always a good idea to have a look at these dialog boxes to see what they're asking for. In this case, it's asking you to choose the profile you'd like to rotate (Figure 4-102).

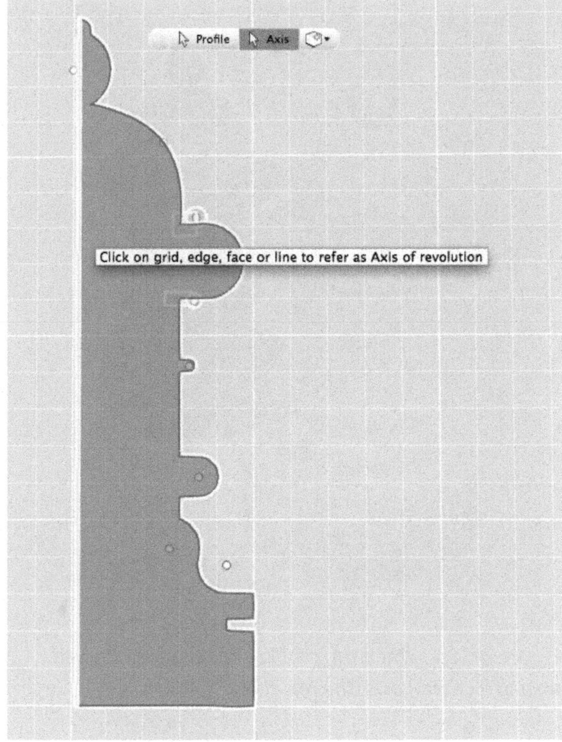

Figure 4-102 *Choose the profile to rotate*

You'll see your selected profile highlighted in blue. Next, select the Axis button in the pop-up menu dialog box (Figure 4-103). This is how you choose the line that will be the axis.

In the dialog box, type the amount that you'd like to rotate the profile. In most cases (this one included), you'll want to rotate through 360 degrees.

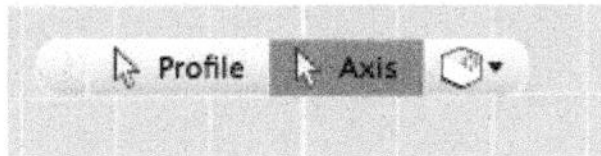

Figure 4-103 *Select the Axis button to rotate the profile on its axis*

Figure 4-104 shows the final result after revolving a profile on an attached axis. The holes were added after the revolve.

Figure 4-104 *When the profile of a shape is revolved around a central axis, the result is a 3D object*

You can also revolve a shape around an axis that isn't attached to the profile. For example, see Figure 4-105 and Figure 4-106.

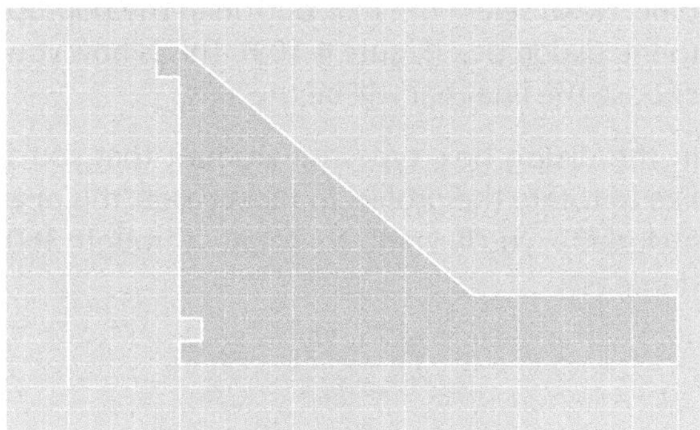

Figure 4-105 *Note that the axis is detached and is several units to the left of the profile*

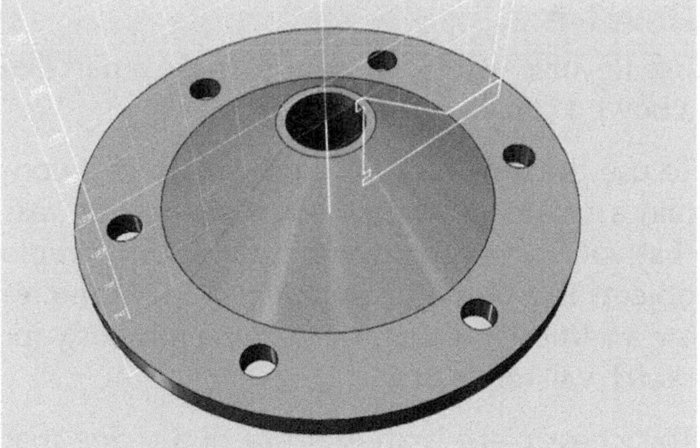

Figure 4-106 *Revolving a profile around a detached axis creates an opening in the final shape*

Loft

Use Loft when you want to create a complex shape based on multiple profiles or sections.

Lofting is the favored choice for product designers, especially those looking to make ergonomic shapes. For example, let's say you need a handle for an ice axe, which is a rather complex shape (Figure 4-107).

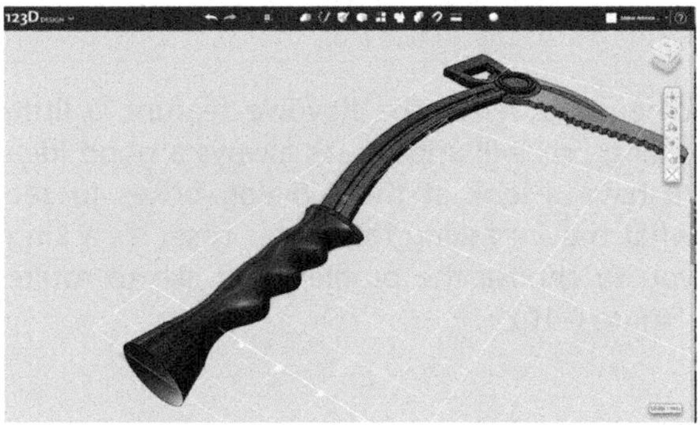

Figure 4-107 *How to make a complex shape by combining profiles*

This operation is called lofting because you literally loft one sketch up to another. It may sound complex, but it's really quite simple. You start with a profile for the beginning of the loft, and then choose a second profile that will guide the first profile to the place you want it to go. Lofting is also a chance to use some projected geometry.

To get started, you'll need to create a series of sketches based roughly on an original. I suggest starting with a basic shape by copying and pasting it to set up your planes. You can do this by using Ctrl-C and Ctrl-V on any closed sketch (Figure 4-108).

Figure 4-108 *Start with a circle*

Be sure that your closed sketch doesn't have overlapping lines. When a loft doesn't work, it's usually because the sketch has lines that are overlapping or aren't connected (and therefore the sketch isn't closed).

Let's try it:

Sketch a circle. Then copy and paste your circle a few times. With each paste, drag the circle vertically to somewhere else in the sketching space (Figure 4-109).

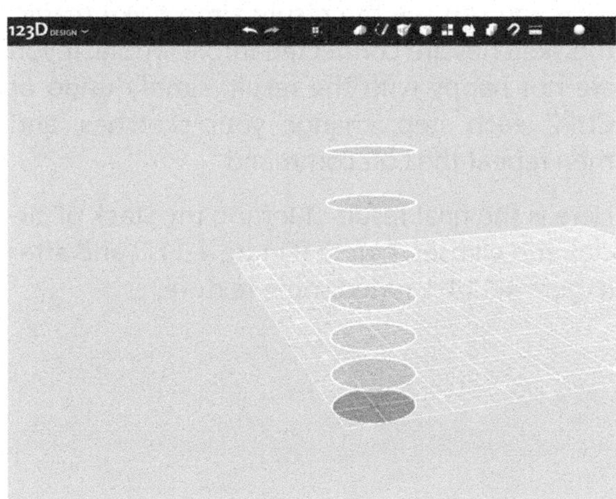

Figure 4-109 *Copy and paste several sketches of a circle vertically in the main plane*

For this particular example, I offset each circle with an ellipse, giving the shape the characteristics of a hand grip (Figure 4-110).

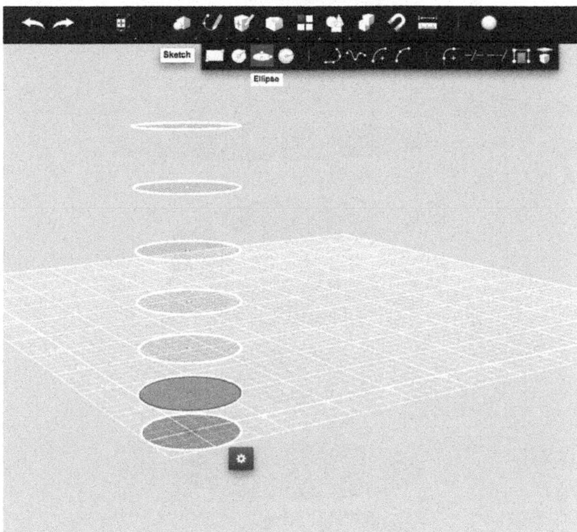

Figure 4-110 *Sketches of circles alternating with ellipses*

In this scenario, as in most lofting scenarios, you're going to loft more than one sketch to another. To do this, hold down the Shift key and select each sketch, in order. In other words, start at one end and work your way toward the other (Figure 4-111).

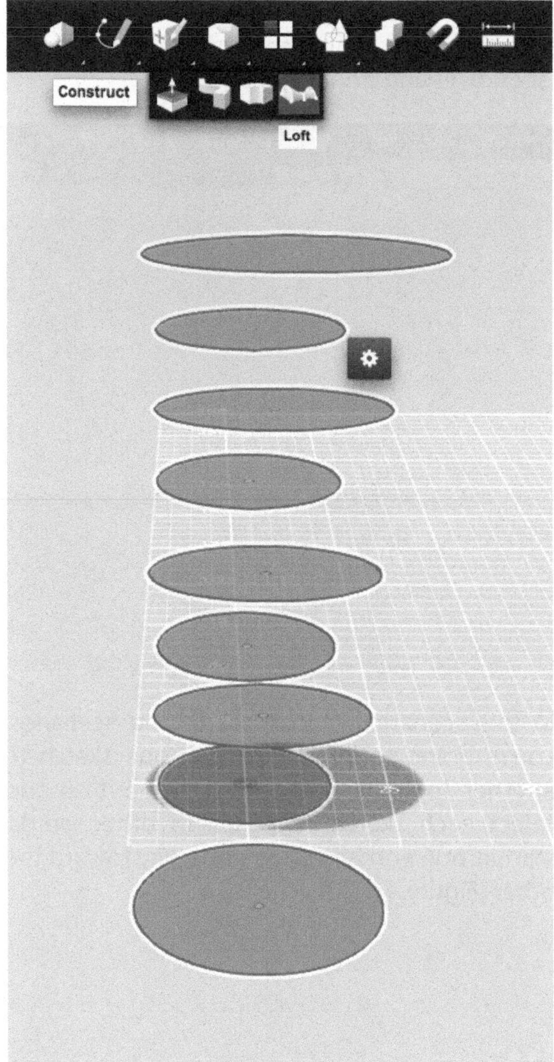

Figure 4-111 *Hold down the Shift key and select each sketch in order, either from top to bottom or bottom to top*

With all the shapes selected, click on the Loft tool from the pop-up menu or the Construct toolbar (Figure 4-112).

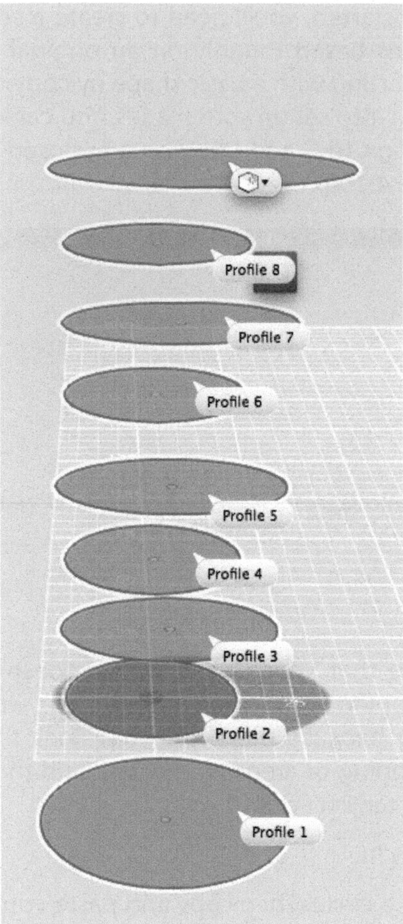

Figure 4-112 *Click on the Loft tool after all the shapes are selected*

Now you can see the result: almost like magic, the sketches are connected into a whole. If you are not happy with the result, simply undo or Ctrl-Z each step, change your sketches, and then repeat the Loft command.

Here is the final result of lofting my stack of circles and ellipses, before (Figure 4-113) and after (Figure 4-114) I added some texture.

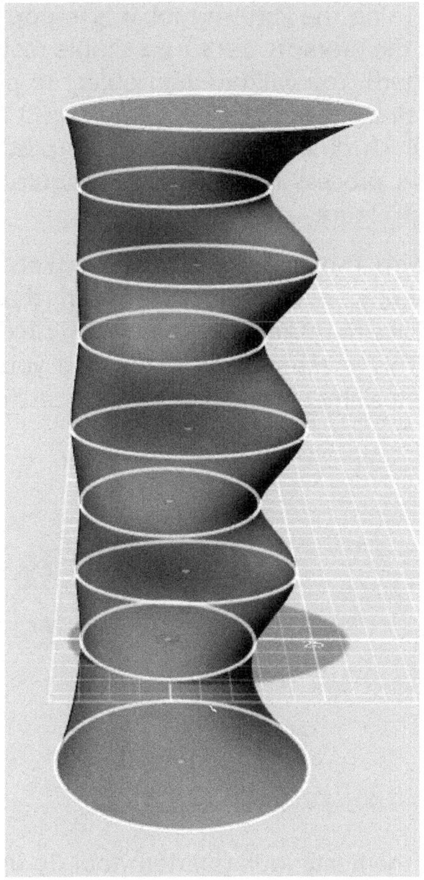

Figure 4-113 *The lofted stack of circles and ellipses*

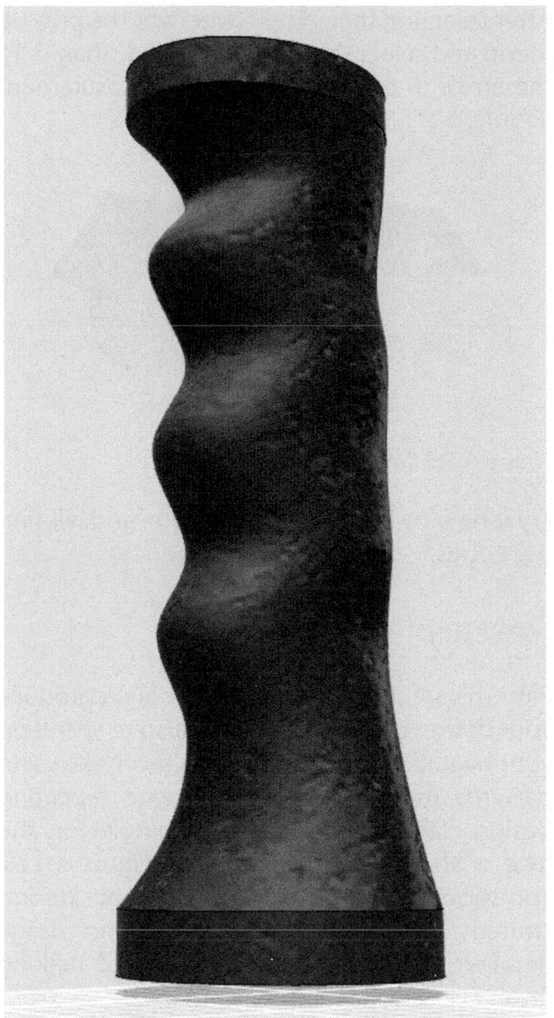

Figure 4-114 *With the addition of texture, the lofted object begins to resemble a handgrip*

Modify

Use the tools in Modify to build out, refine, and add gross details to sketches and models.

Fillet

As with the sketch fillet tool earlier, fillets should always be added last, as they can turn a boring straightforward design into an organic masterpiece.

To use a fillet, simply select any edge. Edges are marked with black lines tracing the object.

After selecting the edge, hover over the pop-up menu and select the fillet command. Than drag the arrow in or out or enter in a measurement (Figure 4-115).

Figure 4-115 *Select an edge... and fillet it!*

Try it now by building a simple box and filleting the edges.

Pattern

Patterns are a great way to not only reproduce your design multiple times but also to speed up your overall design process. For example, using patterns makes it easier to create repeating design elements, such as the shingles on the roof of the birdhouse shown in Figure 4-116. You should start using patterns in your design strategy, especially when creating large detailed objects. It's a great way of making fewer clicks while designing.

Figure 4-116 *A simple birdhouse, with a patterned shingle roof*

When using the Pattern tool, it is important to follow the prompts. Let's try a simple rectangular pattern. You will need an object to pattern and a straight-edge corner on an object to dictate which direction you want the pattern to go. This process is illustrated in Figure 4-117 through Figure 4-122

Start with a solid model or a closed sketch. You must know what kind of pattern that you would like to make first: rectangular, circular, or path. This is essential to make sure you have the right elements in place to be successful.

Figure 4-117 *Start with solid model*

Select the Rectangular pattern tool. To use this tool you must have a straight line (horizontal or vertical on your solid or sketch). If you do not have this element, simply create a polyline sketch on your grid (not touching the element you want to pattern) for your pattern to follow. If you make a horizontal line your pattern will move in a horizontal nature and vice versa.

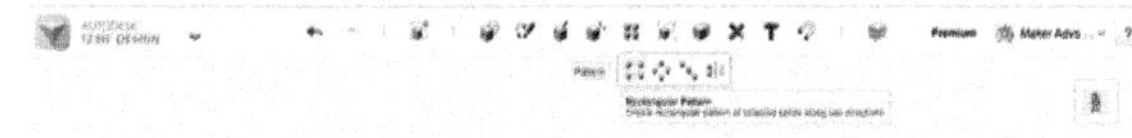

Figure 4-118 *Choose the Rectangular pattern tool*

Figure 4-119 *Follow the pop up menu and select your object*

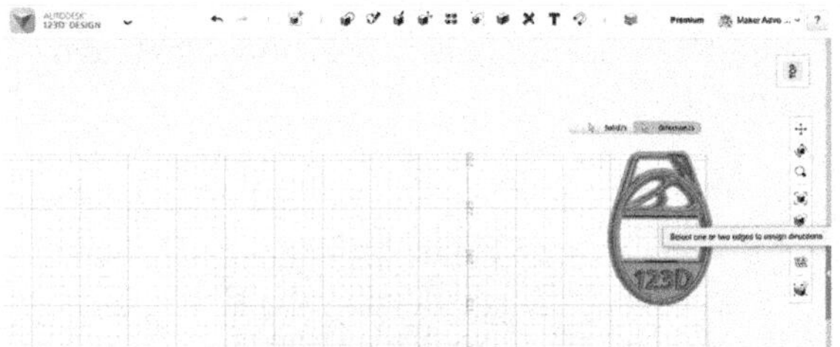

Figure 4-120 *Then select a straight line on that object for which your pattern will follow*

Enter the number of clones you would like to create in the dialogue box; some of them may overlap. Pull the white arrow in the direction you would like to go to spread them out. You can also insert a measurement to determine the distance between the two objects.

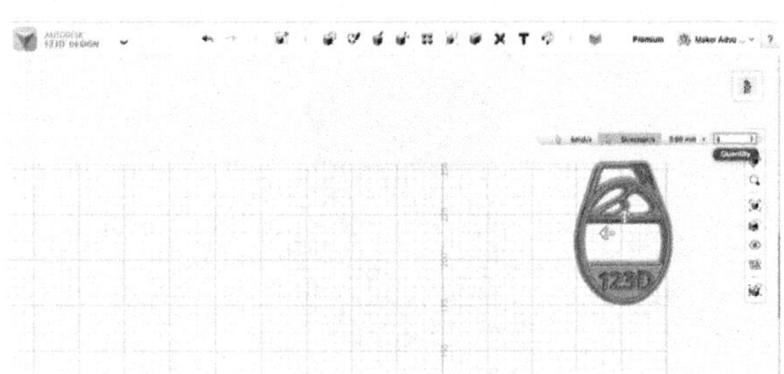

Figure 4-121 *Type the number of things you would like to create with your pattern*

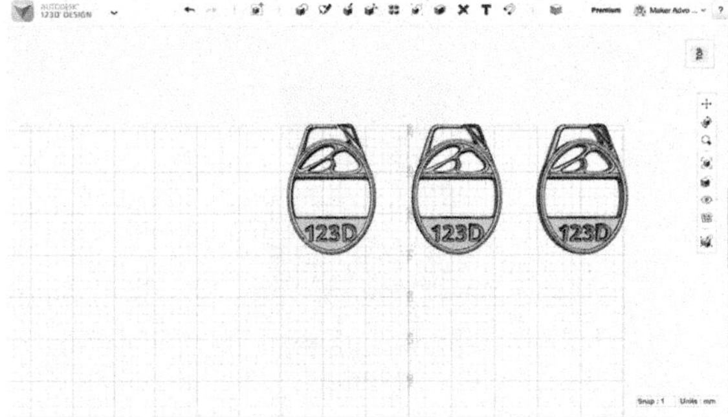

Figure 4-122 *Boom, you have just patterned a solid object*

Another type of patterning that can be useful is the Circle Pattern tool, which is helpful for modifying a round object.

Start by sketching the object that you would like to add or subtract from your object (Figure 4-123). You will need to move your sketch to the area where you would like the feature placed. You can do this by using the Move command (Figure 4-124).

Figure 4-123 *Sketch a circle*

Figure 4-124 *Move your sketch to the desired area on your model*

You must perform an action like an extrusion before you can create a circle pattern.

Select the place or the object that is a result of your action, such as selecting the inside wall of where you subtracted from your object (Figure 4-125).

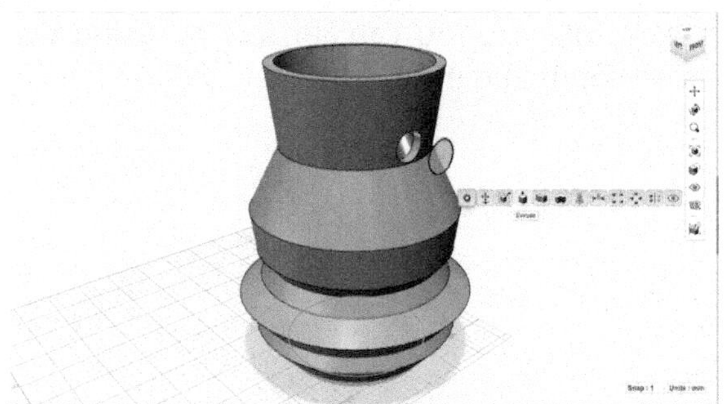

Figure 4-125 *Create a feature with your sketch, in this case we have made a hole*

You may select any circle to create the pattern from (Figure 4-126), but know that it will only create the pattern around the diameter of the circle that you choose (Figure 4-127).

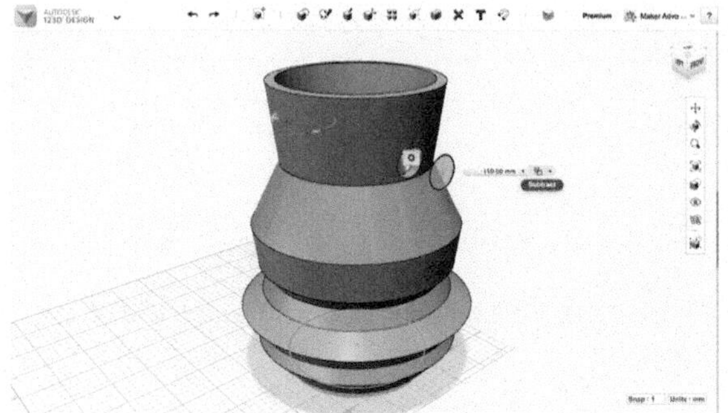

Figure 4-126 *Select Circular pattern from the Pattern tool menu*

Figure 4-127 *Use the pop up menu to choose first your feature (in this case the inside of that hole you just made), then choose an axis to move it around*

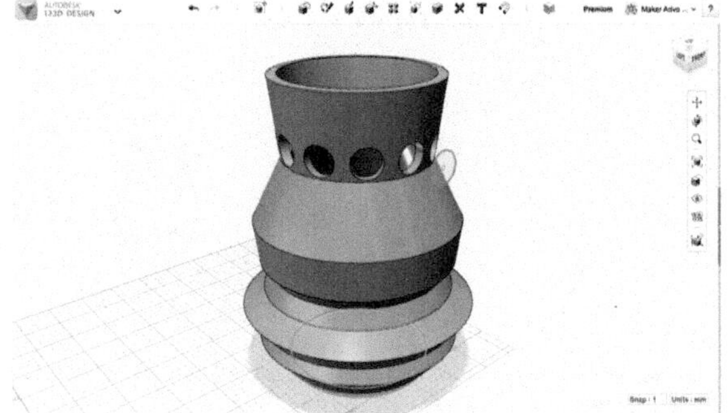

Figure 4-128 *Here we see the finished result of the Circular Pattern tool*

There are other types of patterns as well, including patterns that can follow a path. In addition, there is a Mirror feature that is helpful to make an object that must be symmetrical on both sides, such as the guitar body. It would be very difficult to get both sides of the guitar to match if you sketched out the entire shape.

Instead, simply create half the guitar body and then mirror that shape. This way they will match perfectly.

Grouping

Because 123D Design has no design history tree, the Group tool is essential if you are making a larger, more complicated design that has multiple parts. When building a creature of multiple parts, this offers a great way to be able to select the body and the feature you are working on independently of the rest of the object. You can ungroup parts when you're done working with them as a group, or regroup some of them into new groups with other parts.

Combine

Just as it sounds, the Combine tool combines multiple objects. Unlike the Group tool, however, you *cannot* uncombine objects.

Combine is what is often referred to as a *Boolean operation*: an operation in which the objects you are working with have only two possible values. In the case of the Combine tool, your objects are either all separate, or all combined—there is no middle ground.

Text

This tool is self-explanatory, and a fun means to customize your creations.

To use the Text tool, select the text icon. Then select the surface on which you would like to place your text. You may choose from a large library of fonts, but some fonts will not be able to be used by this program.

Let's take the example of an Arduino enclosure (Figure 4-129). You can see that along one side, I've used the Text tool to create letter shapes, and then various other tools to raise the letters off the surface of the enclosure, and give them texture.

Figure 4-129 *The design for this Arduino enclosure includes text that has been extruded in order to project slightly from the surface of the box*

Snap

It can be difficult to position models atop or alongside each other perfectly by moving them around freehand. This is where the Snap tool comes in.

Snap is located on the right hand side of the tool bar: it is the icon shaped like a U-magnet. Select Snap, then select the model you wish to move, and then select the model you wish to move it to.

Let's say you have a pyramid and a cube on the main plane, and you wish to center the pyramid perfectly atop a cube:

Click the Snap tool.

Then, click the bottom face of the pyramid.

Next, click the top face of the cube.

The two models will snap together, along the faces selected.

Material

To the right of the action tools, there is a patterned cube icon. Click it to bring up a menu of surface material choices—wood, metals, plastics, and so on—that you can apply to your shapes. You can also dial up and apply different colors using the color selector interface to the right of the materials selection.

Reading about the different tools in 123D Design can be educational, but you've got to get your hands dirty if you want to master them. Try out the tutorials that follow so you can get an idea of just how powerful 3D design can be.

If you have trouble remembering where the different tools are, use the preceding sections to refresh your memory.

Tutorial: The Key to Success

For this tutorial, we're going to use all we've learned in this chapter to produce a simple key. It may not seem like much, but the humble key offers examples of arcs, circles, polylines, extrusion, fillets, and so many other things we learned in this chapter.

1. Open up a new project.
2. Sketch two circles (Figure 4-130).

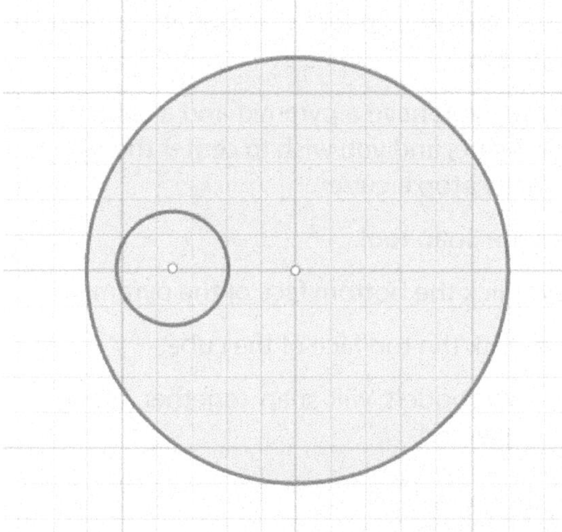

Figure 4-130 *Two circles sketched on the main plane*

3. Use the Polyline tool to sketch the teeth of the key. Make sure to close the sketch (Figure 4-131).

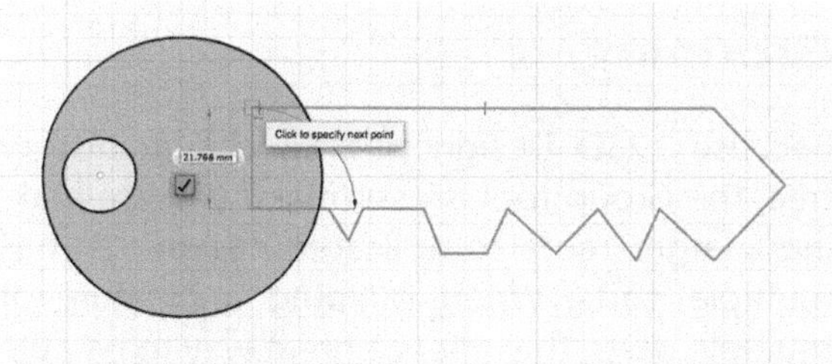

Figure 4-131 *Teeth of a key created with the Polyline tool*

4. Select multiple sketches to create a highlighted key shape. You can do this by holding the Shift key as you select each piece.
5. Hover over the gear icon and select the Extrude feature. Extrude your key by pulling up on the arrow (Figure 4-132).

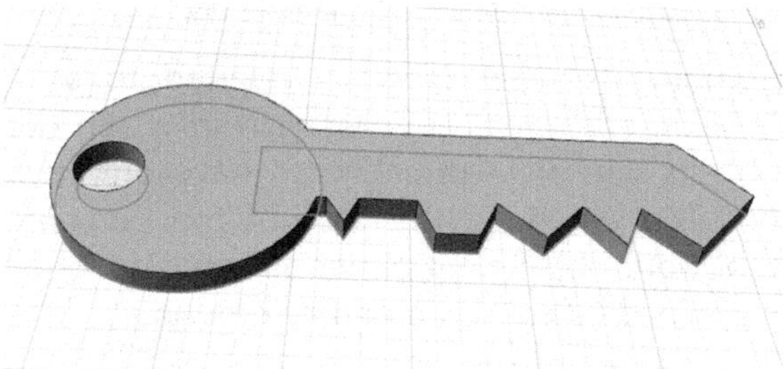

Figure 4-132 *The circles and polyline, extruded into a single key shape*

6. Save and close the project, if you desire.

Great job!!! You are well on your way to global domination.

Tutorial: Guitar Body

Like a key, a guitar uses many of the concepts in this chapter, including complex splines, extrusion, and mirroring.

1. Open up a new project.
2. Use the Spline tool to create the shape shown in Figure 4-133. Don't hit the checkbox yet!

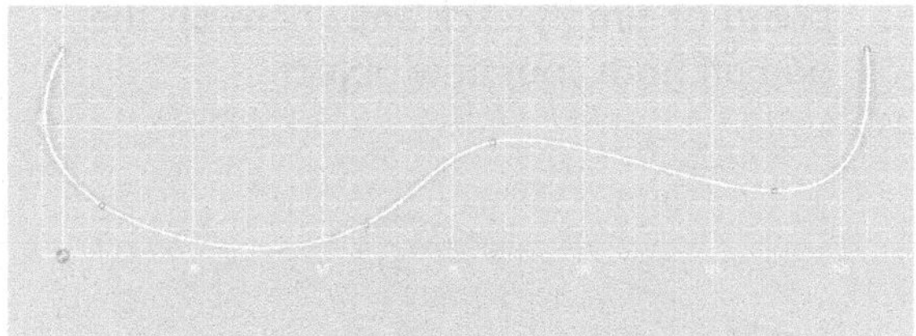

Figure 4-133 *Use the Spline tool to create the two curves along one side of a guitar body*

3. While still in the Spline tool, hover up to the Polyline tool. Your sketch will glow green.
4. Select one end of the spline and draw a line to the other end to close the sketch.
5. Now with your sketch closed, you can refine the shape by selecting and dragging the white anchor points up, down, and side to side (Figure 4-134).

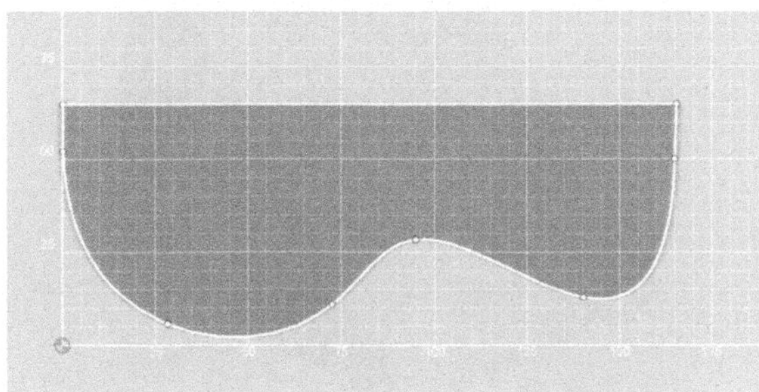

Figure 4-134 *Adjust the curves by clicking and dragging on the white anchor points along the spline*

6. Once you've got the curves where you want them, use the Extrude tool in the Construct menu to bring your design from 2D into 3D.
7. Half a guitar body won't do you much good, so let's use the Mirror command, from the Pattern menu, to add the other half (Figure 4-135).

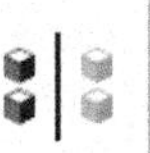

Figure 4-135 *The Mirror tool is ideal for creating perfect symmetry in a design*

8. When it asks you to select a solid, be sure you are selecting the entire body: click and drag over the entire object. Once selected, it will prompt you to choose a mirror plane; select the flat plane with no curves. It will disappear once you have mirrored the object (Figure 4-136).

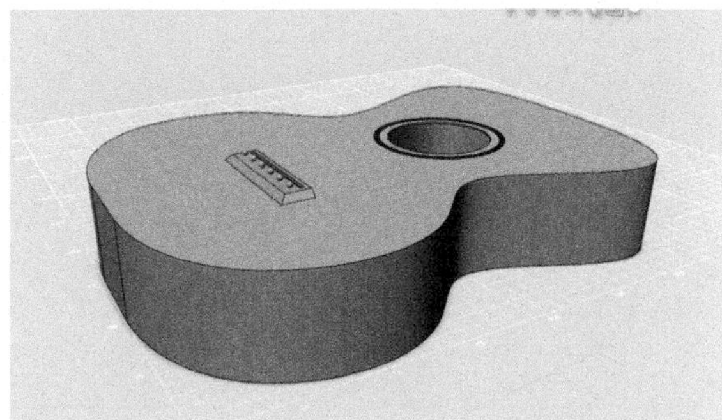

Figure 4-136 *Extruding turns a 2D shape into a 3D object, and then mirroring creates two connected halves with perfectly symmetrical design features*

Tutorial: Revolving a Model

Open the Sketch action toolkit (Figure 4-137) and select the Rectangle tool (Figure 4-138).

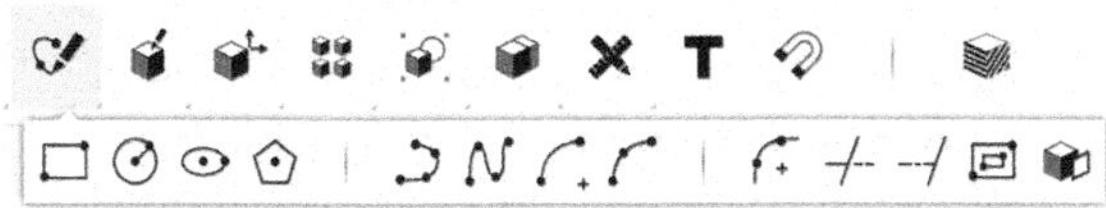

Figure 4-137 *The Sketch action toolkit*

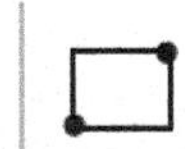

Figure 4-138 *The Rectangle tool*

This final tutorial combines many of the tools of the chapter, along with the Rotation command, to create perfectly symmetrical objects.

1. A box will pop up prompting you to select where you would like to sketch. Click on the only thing on the screen: the big blue grid, aka the main plane.
2. Click again to lock down the first corner of the rectangle. Then move your mouse and click again to complete your rectangle (Figure 4-139). Don't worry right now about the size of the rectangle; we'll go over that a little later.

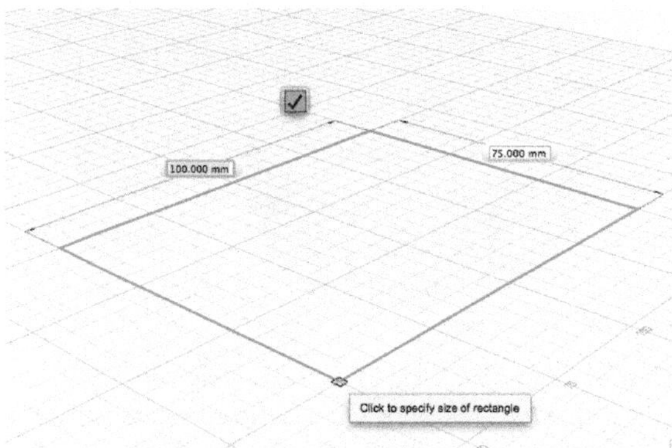

Figure 4-139 *Complete the rectangle*

3. Hit the Escape key on your keyboard to leave the Rectangle tool.
4. Click in the middle of your new rectangle. It should glow blue.
5. Hover your arrow cursor over the gear icon and select the Revolve tool from the pop-out menu. (If these tools look familiar, they should: they're a selection of the same tools in the action tools menu.)
6. You'll see a pop-up menu with two options: Profile and Axis. Click on Axis.
7. Select one side of your square by clicking on it (you can choose the top, one of the sides, or the bottom).
8. To revolve your sketch into a solid, click and drag on the small circle that pops up on the diameter of the large circle (shown in Figure 4-140). Stop where desired. Hit Enter/Return on your keyboard or simply click anywhere on the grid off from your new object.

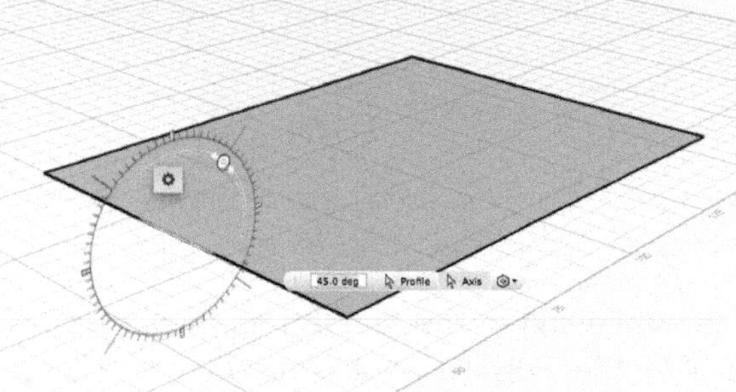

Figure 4-140 *Use the dialogue pop up to rotate your object*

Presto! You've created a 3D model that you can now print, alter, add materials to, or just stare at affectionately.

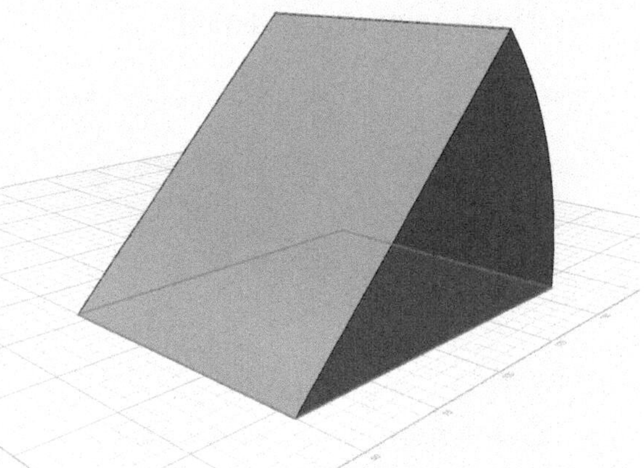

Figure 4-141 *Finished result of only rotating an object 35 degrees*

Challenge: Build a Birdhouse

Now that you've clicked around and gotten familiar with the action tools and other menu options in the 123D workspace, here's a simple exercise to get your 3D design feet wet: re-create the birdhouse shown in Figure 4-142.

Figure 4-142 *How many design processes can you see in this simple birdhouse?*

Remember, it's really just two shapes that are manipulated: a box and a cylinder. Be brave and experiment!

What's Next?

Now that you've got the basics down, what's next with 123D Design?

Lots! This is where the fun starts. You can use the knowledge you've gained in this chapter to make some extremely complicated objects. Remember, there's more than one way to do things in CAD!

If it seems like there are too many tools and techniques to master, don't give up. Over time, you'll get to know the tools. You'll develop your own little tricks and discover new ways to do things. And because each CADer does things a little differently, there's a lot of opportunity to learn from others.

For inspiration, check out Figures 4-143, 4-144, and 4-145, which show a few of the impressive builds that 123D Design users have accomplished. See what you can come up with!

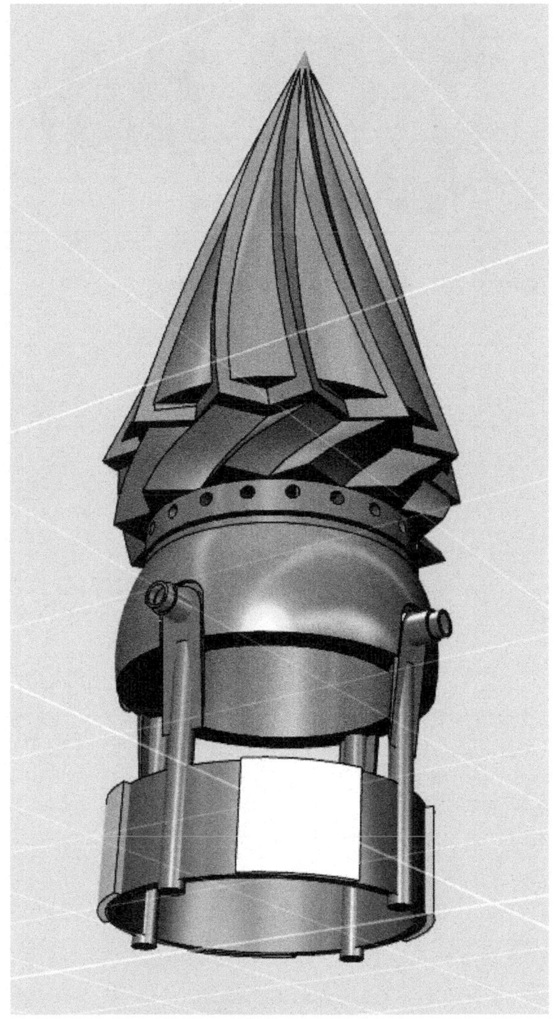

Figure 4-143 *A steampunk drill*

Figure 4-144 *A set of headphones*

Figure 4-145 *Anything you can imagine!*

You can learn more by visiting Autodesk 123D's YouTube channel (*https://www.youtube.com/user/123d*).

123D Make

5

123D Make is a program that helps you turn a 3D model into a real-world object. Unlike printing with a 3D printer, however, 123D Make turns all the parts of the model into a 2D plan: shapes that can be printed on and sliced from flat materials including wood, cardboard, plastics, and metal, and then assembled. This can be a great way to prototype designs. This program is best used with a laser cutter or ShopBot-style CNC machine.

After printing, you can use a laser cutter to slice the shapes out of your material by hand, or you can have them cut and delivered by a third-party service.

123D Make also creates an animated set of assembly instructions, based on the construction method you choose while preparing your model in the app.

123D Make is available for Mac, PC, iPhone, and iPad.

Getting Started

To get started with 123D Make, download the app from 123Dapp.com (Figure 5-1) and install it on your preferred platform.

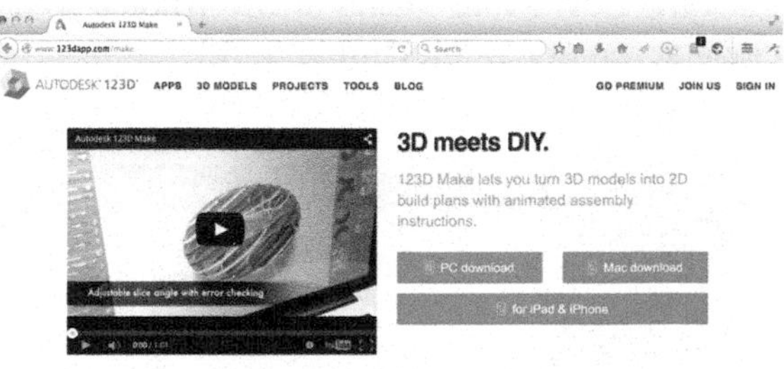

Figure 5-1 *Download 123D Make from 123Dapp.com*

1. When you first open 123D Make, you are presented with several introductory slides that introduce the app's different tools. Take a few seconds to click through these slides, and you'll discover that many of them are similar to other apps in the 123D family.
2. Close the slideshow.
3. Now you are in the app's main workspace (Figure 5-2): a blank screen, with a drop-down menu and tools for importing 3D models to the upper left, and a sign-in button at the upper right. Go ahead and sign in; this should be your first step when using any 123D program.

Figure 5-2 *The 123D Make workspace*

Drop-Down Menu

Click the downward-facing arrow inside a small circle at the upper left (Figure 5-3).

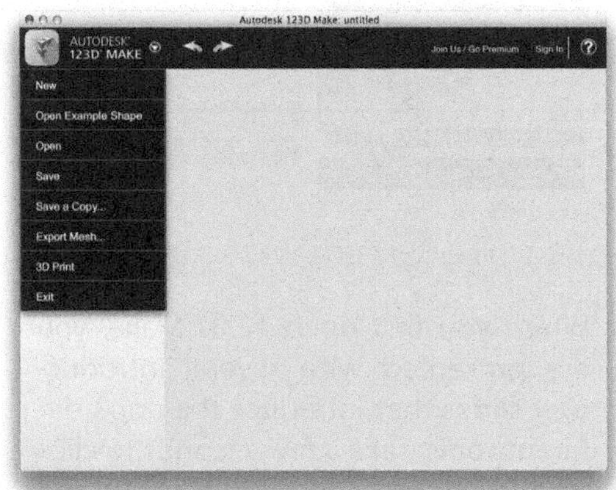

Figure 5-3 *The 123D Make main menu*

A menu will drop down that offers basic file commands; let's take a look at each of these.

New

Creates a new project.

Open Example Shape

Offers a selection of preexisiting shapes such as a car, a flying saucer, a rhinoceros head, and more (Figure 5-4). These are fun and easy to play with, or to use as the basis for a new model. I have used the rocket shape to make many projects, such as this one: *http://www.instructables.com/id/Rocket-Bookshelf/*.

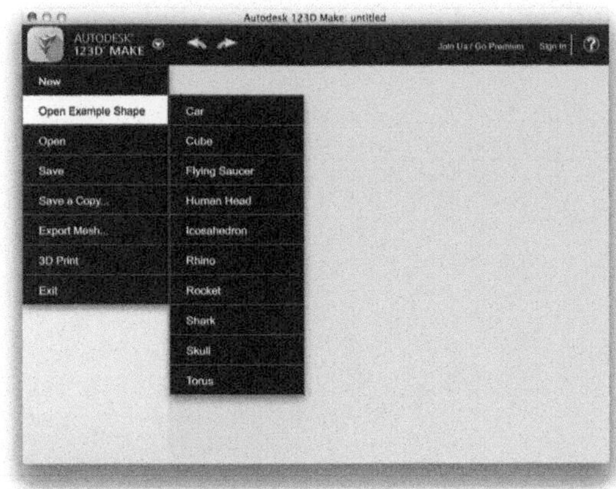

Figure 5-4 *Example shapes make it easier to get started on a new model*

Open

Open previous projects that you have made in 123D Make, or a project from the gallery (Figure 5-5), created by someone else.

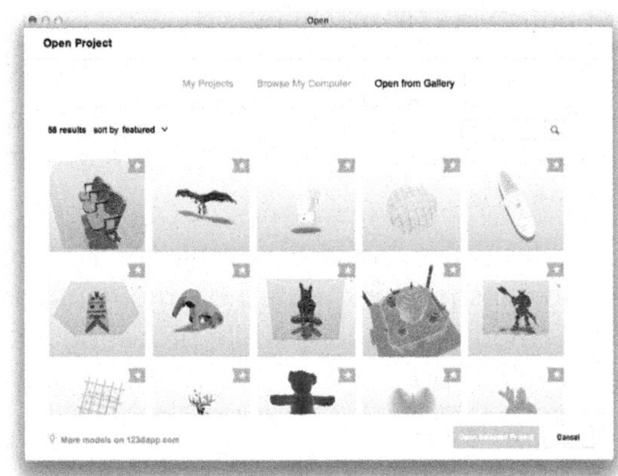

Figure 5-5 *Open a model from the 123D gallery*

Save

Click on Save (Figure 5-6) and you will see two options.

The Save To My Projects option saves a file that holds the settings (not the cut file) that you have used in 123D Make to your 123D account.

The Save To My Computer option saves the 123D Make file to your own computer. It is a best practice to set up a project folder that holds the three essential elements of your project: your 3D model, your Make file, and your cut file.

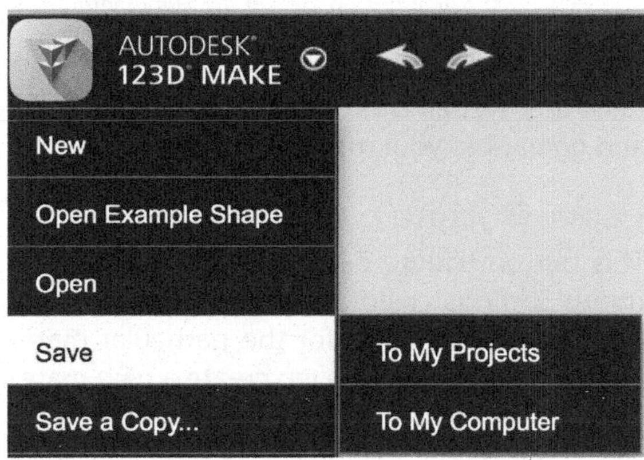

Figure 5-6 *Save project options in 123D Make*

Save a Copy...

This option (Figure 5-7) saves a project in a separate file from the original project file. Useful for trying different ways of manufacturing or different material sizes.

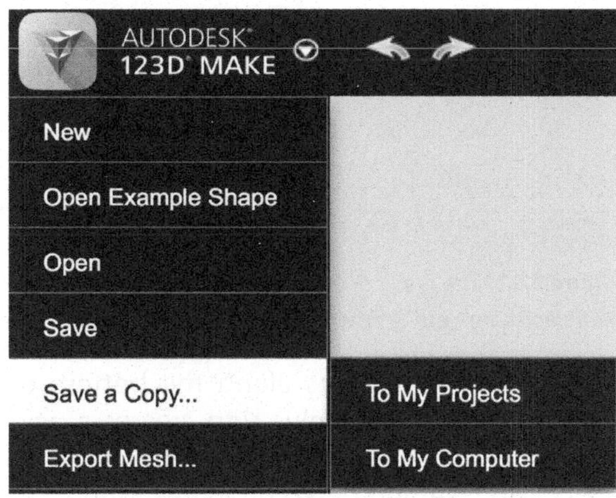

Figure 5-7 *Save copy options in 123D Make*

Export Mesh

One of my favorite features, Export Mesh (Figure 5-8) allows you to save and export a mesh of your sliced or paneled model for 3D printing.

I use this a lot when trying to achieve a more geometric look to a design, as opposed to designing it all in flat planes.

Figure 5-8 *The 123D Export Mesh menu*

3D Print

Similar to Export Mesh, with the exception that this command sends your file directly to the Autodesk 3D print utility that sits within Meshmixer, where it is stitched up and made watertight. You must have Meshmixer installed in order for this command to work. If you don't, you will be prompted to download the program.

Exit

Quits 123D Make.

Model Menu

This menu (Figure 5-9) appears right below the 123D Make logo and file menu, along the left-hand side of the workspace. It contains three buttons.

Import

Allows you to open any STL or OBJ file from your desktop.

Eyeball Button

Located to the right of the Import button, this feature shows you the original model after you have made alterations to it.

Rotate Arrow

Found below the eyeball button. This rotates the model on the *z*-axis.

Figure 5-9 *The Model menu*

Manufacturing Settings Menu

Once you have a model open in the workspace, the Manufacturing Settings menu (Figure 5-10) appears below the Model menu.

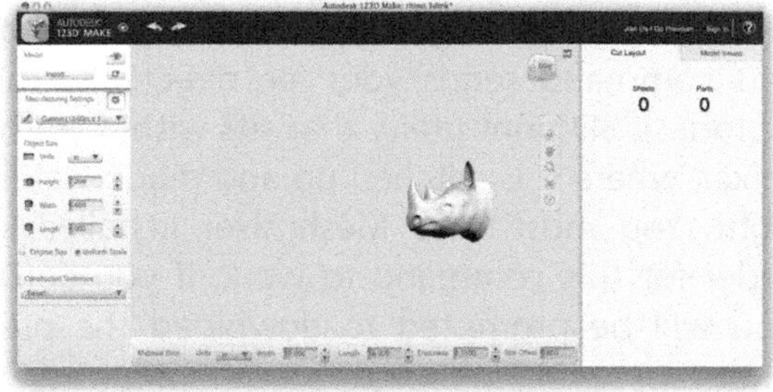

Figure 5-10 *The Manufacturing Settings menu in 123D Make*

It includes a number of tools that allow you to customize the size of your model, the size of the flat material you intend to use, and more.

Gear Button

The gear button (Figure 5-11) contains various manufacturing settings that appear along the bottom and right side of the workspace.

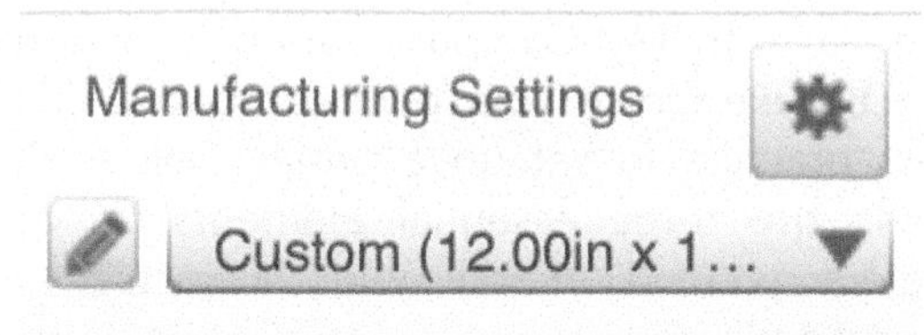

Figure 5-11 *The gear button, part of the Manufacturing Settings menu*

Click the eyeball button to close these panels and go back to your model.

Pencil Button

This button (Figure 5-12) opens the Manufacturing Settings custom dialog box, where you can enter the settings for the particular material you plan to use, or even create a new material and related settings.

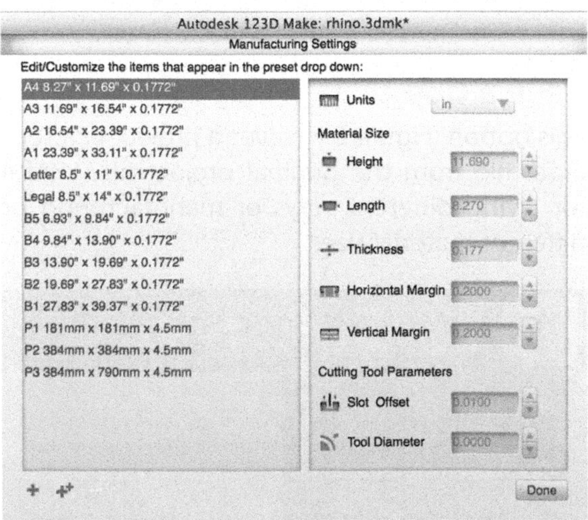

Figure 5-12 *The result of clicking the Pencil button in the Manufacturing Settings menu*

There are three options along the bottom of this menu: The single plus sign creates a new custom setting. The double plus sign creates a new custom setting based on a preexisting setting. The minus sign deletes a selected setting.

If you haven't gotten your flat materials yet, you can make your best guesses, save the 123D Make file, and then alter them as needed once you have your material in hand.

This is important because material thickness can vary enough to prevent joints from fitting, or to create a distorted model.

Custom Menu

The Custom menu is to the left of the Pencil button (Figure 5-13).

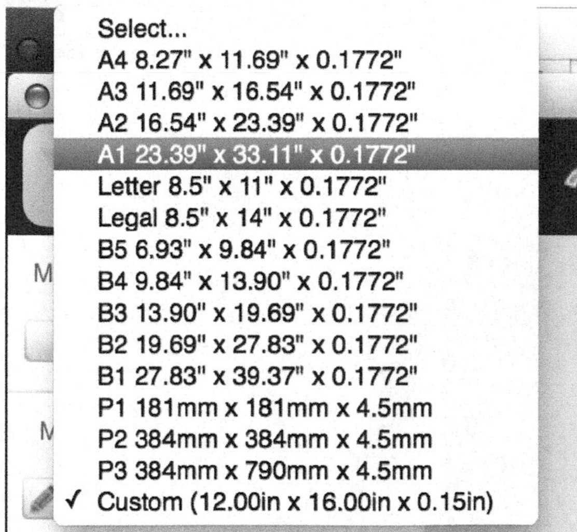

Figure 5-13 *The Custom menu in 123D Make*

Use this drop-down menu to select from a number of preloaded settings for common flat materials, or the custom materials settings that you have created.

Note that after you have entered a custom material under the pencil icon, you must then select it with the drop-down menu to apply it to your part.

Object Size Options

You have a number of choices (Figure 5-14) in setting the size for your model.

Units

Choose inches, feet, millimeters, or centimeters, then enter the height, width, and length you intend for your real-world object. Be aware that if you choose to make an object several feet in length, 123D Make may need more time to process your model.

Original Size

If you designed your object to specific measurements, be sure to check this box!

Uniform Scale

Keeps your design constraints when sizing your object up or down.

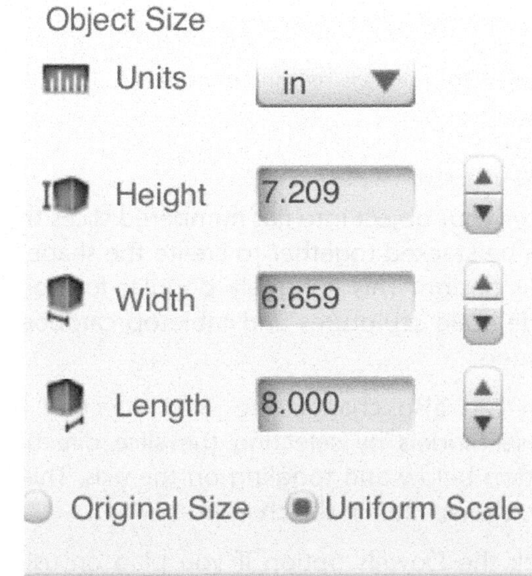

Figure 5-14 *The Object Size menu in 123D Make*

Construction Technique Menu

The Construction Technique drop-down menu (Figure 5-15) offers a variety of options. Note that as you choose particular techniques and change their settings, to the right of your workspace the "sheets" and "parts count" also change. Generally speaking, the more sheets of material you need, the more expensive your object will be. The higher the number of parts, the more difficult, or at least time consuming, it will be to assemble the object.

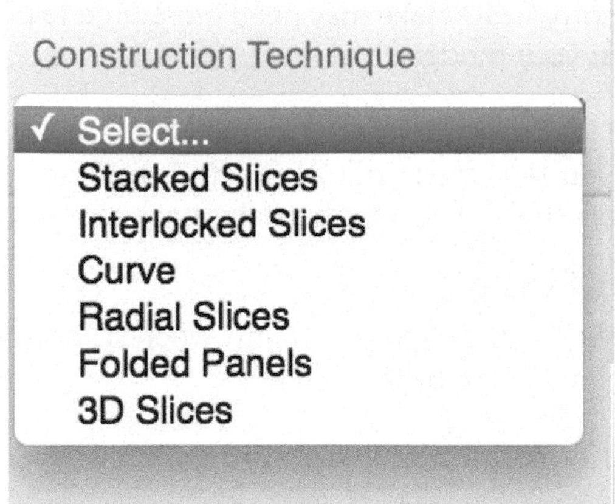

Figure 5-15 *The Construction Technique menu in 123D Make*

Stacked Slices

Slices your object into flat numbered slices that can be stacked together to create the shape of your design. This is hugely popular for doing large scale sculptures and tabletop cardboard models.

You can also change the slice direction for these models by selecting the slice direction button below and toggling on the axis. This is also shown later in the chapter.

Click the Dowels option if you plan on using dowels as supports to pierce and stack the pieces of your object. Dowel shapes can be Square, Pencil, Round, Cross, Horizontal Slot, and Vertical Slot.

Be sure to enter the correct diameter for the dowels (or dowel-like objects) you intend to use in construction.

Figure 5-16 through Figure 5-19 show some incredible examples of what people have created with Stacked Slices.

Figure 5-16 *By Jesse Harrington Au*

Figure 5-17 *By Evan Atherton*

Figure 5-18 *By Jesse Harrington Au*

Figure 5-19 *By Instructables member Damonite*

Interlocked Slices

This technique creates a waffle-like pattern as shown in Figure 5-20. It works very well for building larger structures quickly, as well as for building custom shelving. Interlocked Slices seem to do better with more robust shapes such as a cube or human head. Interlocked Slices must be perpendicular—in other words, they need to be at 90-degree angles to each other as seen in Figure 5-21.

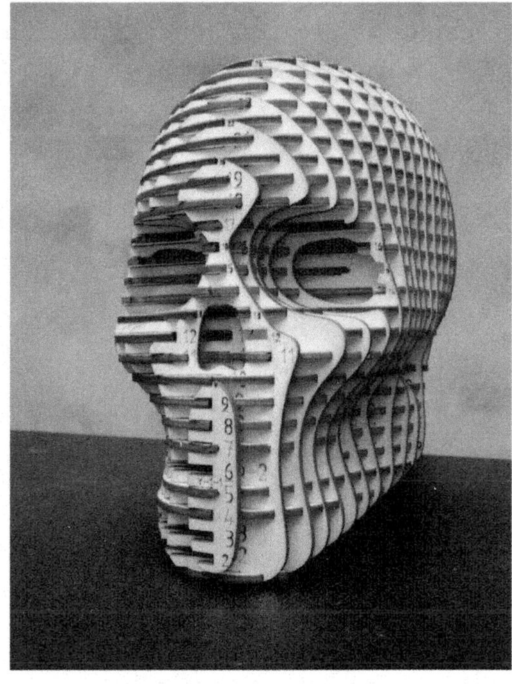

Figure 5-20 *By ScaleFocus.nl*

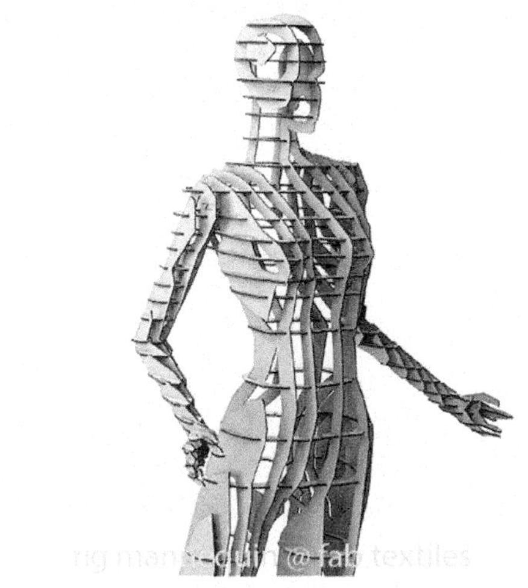

Figure 5-21 *By Fabtextiles.org*

Curve

Curve is one of the coolest ways of slicing your design. It creates a waffle pattern while following along a desired curve. To use the Curve command, select it from the drop-down menu and then select Slice Direction. You will see two axis options as well as a series of blue dots along a line.

As you can see in Figure 5-22, the original form is a lackluster shark.

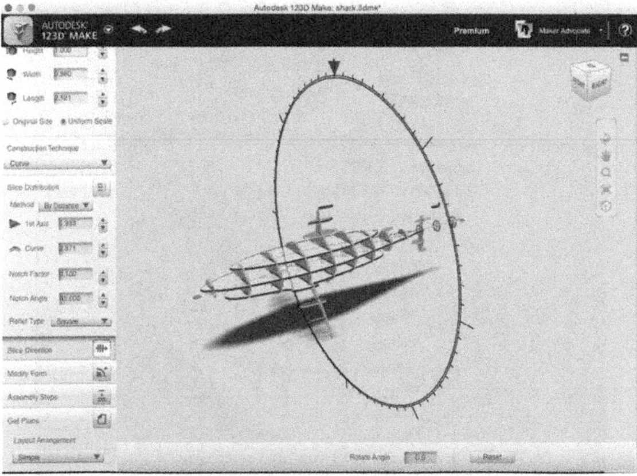

Figure 5-22 *A not-so-great shark*

When we pull the blue arrow on the vertical axis about 90 degrees, the design starts to look a little more like a shark (Figure 5-23).

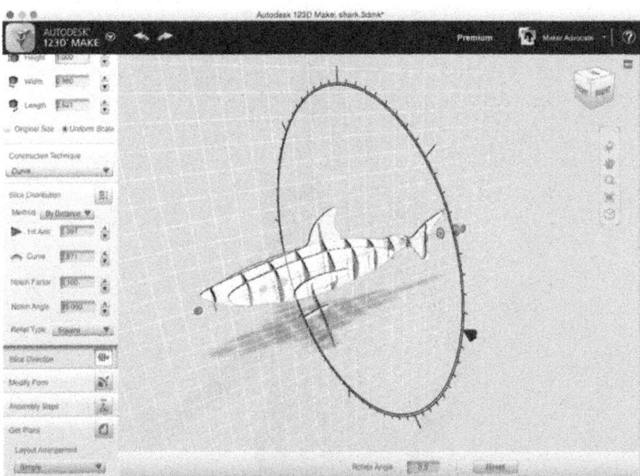

Figure 5-23 *The curve helps create a more shark-like shape*

By grabbing the blue circles and moving them to follow the curve of the shark's back, the design becomes much more memorable (Figure 5-24).

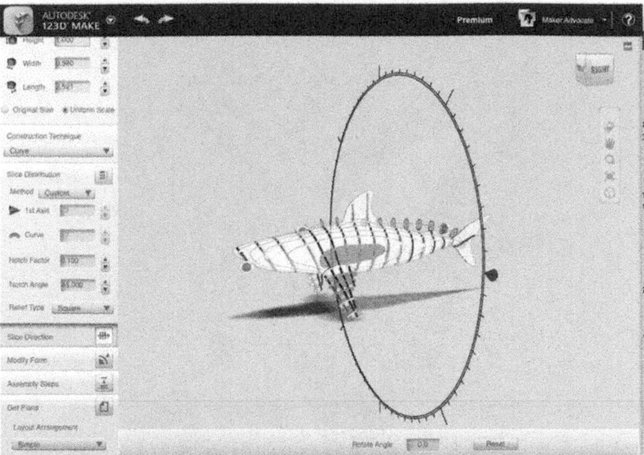

Figure 5-24 *Now we're getting somewhere*

Figure 5-25 shows a rad example of what is possible.

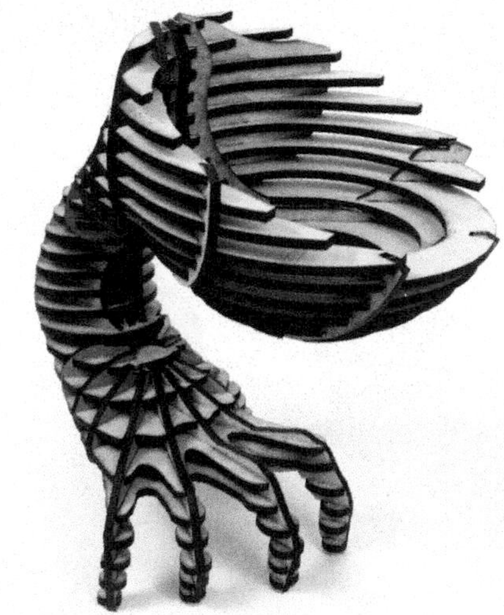

Figure 5-25 *By Instructables member Penfoldplant*

Radial Slices

The Radial Slices option is great for projects that are largely cylindrical in shape, such as my rocket ship bookshelf (Figure 5-26).

Figure 5-26 *Rocket ship bookshelf by Jesse Harrington Au*

Folded Panels

The Folded Panels option is by far the most versatile of the slicing methods available in 123D Make. Its uses have ranged from creating a 12-foot-tall cardboard Trojan horse, to creative stuffed animals, to 3D prints that have a paneled look. However, if you use Folded Panels to create larger forms, the final object will lack any support structure; that will need to be added manually.

As you can imagine, this is bit trickier than other construction techniques. Luckily, the app intelligently labels each piece like a puzzle to assist you in fabrication. Think "insert z-09 to adjacent piece z-10." Not impossible, but takes a bit of tinkering before you begin. A few examples are shown in Figures 5-27, 5-28, and 5-29.

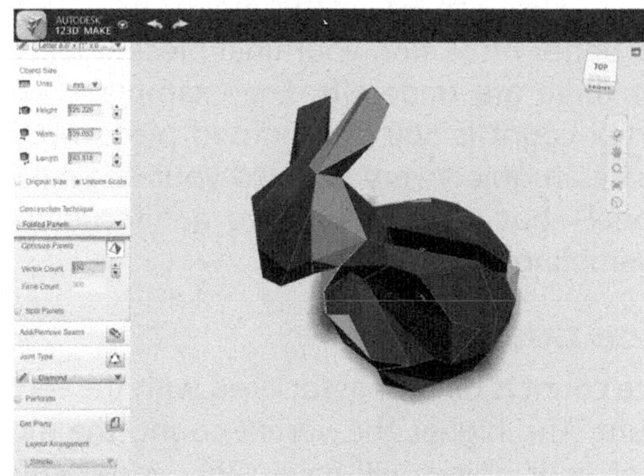

Figure 5-27 *Folded panels before 3D printing*

Figure 5-28 *Printed model created using folded panels*

Figure 5-29 *Another cool example of the folded panel model*

Vertex Count

This establishes the number of edges each individual piece can have. The higher your count, the better resolution your final piece will have, but there are trade-offs. For example, if your vertex count is 500, the printed pieces would create an object very close to your original 3D model—but you would have a difficult time assembling it.

Face Count

Face count is directly associated with the vertex count. The higher the vertex count, the more faces your object will have. Only vertex count can be changed manually.

Add/Remove Seams

This option functions just as it sounds: when in folded panel mode, this command will add or remove seams from your model (Figure 5-30).

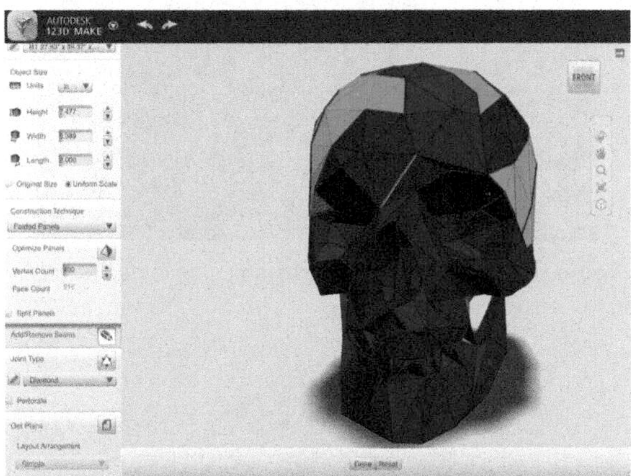

Figure 5-30 *Removing seams from a skull*

Joint Type

Your joint type will determine how you put your pieces together after you cut them out. This is pivotal to your success. You should deeply consider what material you will be using and how you will be cutting out the shapes before making the final decision on joint type. You can always go back and change it, but it might cost you a lot of cutting time and some valuable materials. For instance, if I know that I am using cardboard to construct, I prefer using the rivet technique and then using zip ties in place of rivets. You can also use this method for Metal, unless you are welding; in that case you can use Seam.

Diamond

Creates the shape of each individual piece up to the edge. No real joinery created.

Gear

Creates pieces that have matching teeth that can be interlocked to assemble your model.

Laced

Think shoelace eyelets: this selection creates holes that can be used to zip-tie or lace something together.

Multitab

This selection creates a toothed tab that sticks out from one side and can be press-fit into the other.

Puzzle

Using the Puzzle selection will turn your connections into puzzle pieces. This is not a great way to make a 3D model, but it is a really fun way to make a 2.5D model—in other words, something more flat than truly 3D.

Figure 5-31 shows what the pieces look like.

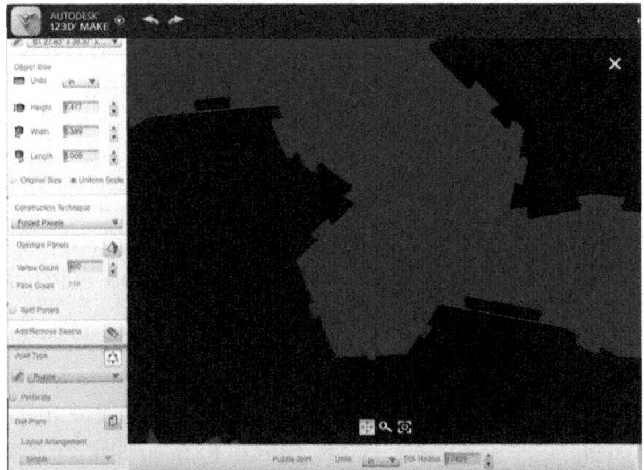

Figure 5-31 *Puzzle pieces*

Rivet

Creates a tab with holes on one side and corresponding holes with no tab on the other so the joint is tight together. Great for using zip ties in cardboard! This is what was used in the Yeti sculpture in Figure 5-32 and also in the large Trojan horse sculpture shown in Figure 5-33.

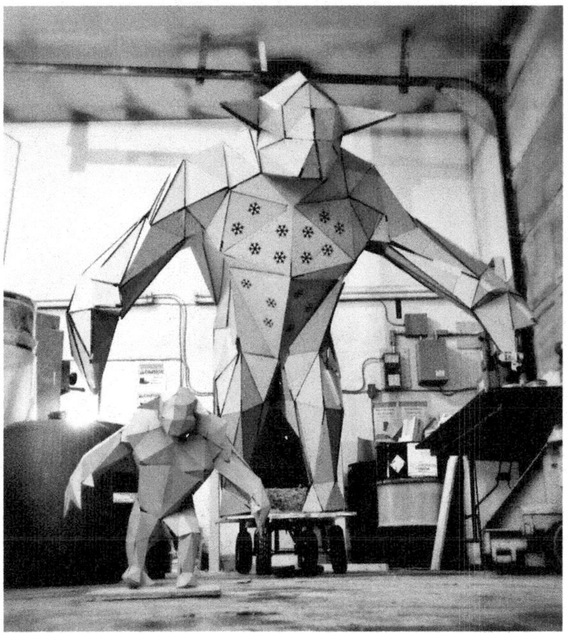

Figure 5-32 *By Yukonstruct*

Figure 5-33 *by Instructables member Penfoldplant*

Seam

Creates extra material around the edge of each shape, called a seam allowance, that makes it possible to sew pieces together without fraying the edge of a fabric. Great for soft goods.

You can laser cut most fabrics, and this is a great way to make a complex form.

Tab

Creates tabs and corresponding slots to assist in locking in.

Ticked

Made for paper, Ticked creates paper tabs around your object that have cut lines built in so you can glue tab to tab.

Tongue

Creates a tab on one side and two slits on the corresponding side to slide the tab through.

Custom

Form a custom joint by selecting this option from the drop-down menu, clicking on the Pencil tool, and entering your desired measurements on the bottom panel. Be sure to label your new joint.

To Perforate or Not?

When you select Folded Panels as your construction technique, the Perforate button appears just below the Joint Type drop-down menu. Check this box to add perforated fold lines to the cut file.

These can be very useful for some harder materials, but the perforations will remain visible in your final object. I have seen this used when water jetting or CNC cutting metal.

You can see it here with the close-up of the horse head (Figure 5-34); the perforations were used to bend the cardboard where we wanted it. This way we could use larger pieces of cardboard and have them bend in the middle as opposed to having to zip-tie every single triangle together.

Figure 5-34 *A perforated horse*

Slice Distribution

Slice Distribution determines how many slices you have in each direction (Figure 5-35). So, if you choose interlocked, curve, or radial, you will have two directions of slicing. You can alter your distribution to add more slices to a direction. If you only choose one slice, you will see that there is only one piece of material in that direction.

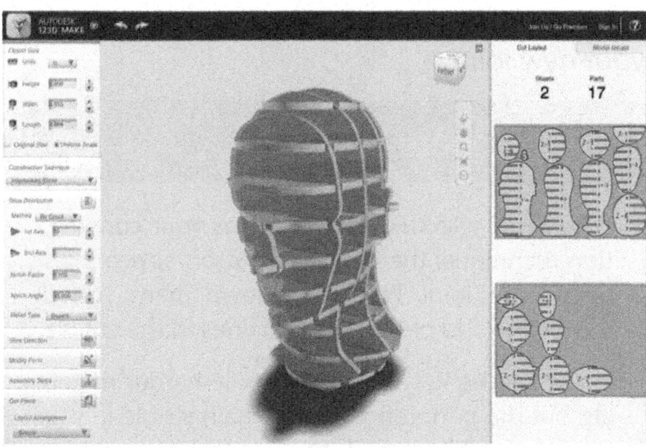

Figure 5-35 *Slicing a head by count*

Choose "By Count" or "By Distance."

1st Axis

First angle of your slices. Toggle up or down to determine the amount of slicing. This will either add pieces or subtract them from your model. Here I have changed the 1st Axis to 20 (Figure 5-36).

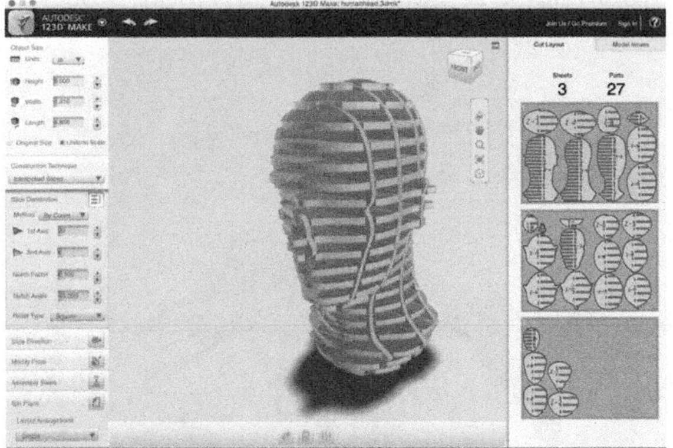

Figure 5-36 *Slicing a head by distance*

2nd Axis

Second angle of your slices. Toggle as above. This time I changed it by measurement and set it at .175, which is about the size of a piece of cardboard between each piece (Figure 5-37).

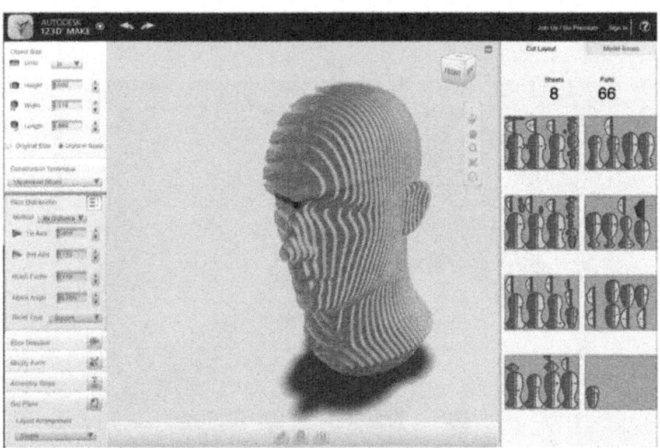

Figure 5-37 *Choosing a second angle of slicing*

Notch Factor

Creates a chamfer (angle cut) on the corners of each finger. This will help you greatly when putting things together in real life. Right angles on laser cut cardboard will look pretty but will make all of your hair fall out when trying to put it together.

Notch Angle
: Determines the angle of the chamfer used at the end of each finger.

Relief Type
: Changes the base of each notch. The only real options here are square or dog bone. Dog bone is useful with any interlocked construction technique, as it allows you to use thicker materials, as well as more advanced manufacturing techniques such as CNC routing. This is because CNC routers have cylindrical cutting bits, and cylinders cannot make nice squared ends. So the dog bone shape allows some room to make sure your pieces that were cut out on a router will sit flush to one another.

Slice Direction

Slice Direction is going to determine which angle your slices will follow. It is a bit tough to explain, so take a look at Figure 5-38 through Figure 5-40. Basically this option determines in which direction(s) you will be slicing your model. The easiest case for this would be stacked slices. If you choose to do stacked slices, 123D Make will automatically slice your model in the vertical axis (i.e., the *z*-axis). When you alter the slice direction, you will be moving from the vertical toward the *x* or *y*, thus creating a more dynamic look to your piece.

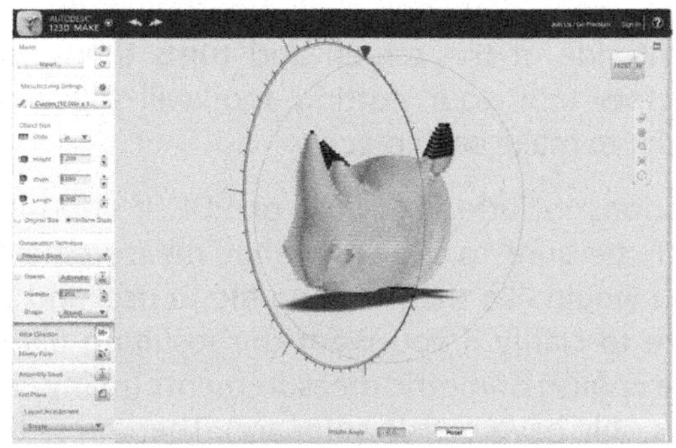

Figure 5-38 *Locking in a slice direction*

You can use this for any slicing option with the exception of folded panels. Folded panels only sit on the surface of your model and do not technically slice your model.

To use Slice Direction, click on the icon next to the command for Slice Direction on the left side menu. Then drag that blue arrow in the desired direction (Figure 5-39).

Figure 5-39 *Drag the blue arrow...*

You can see how it adjusts the direction that it is creating your stacked slices (Figure 5-40).

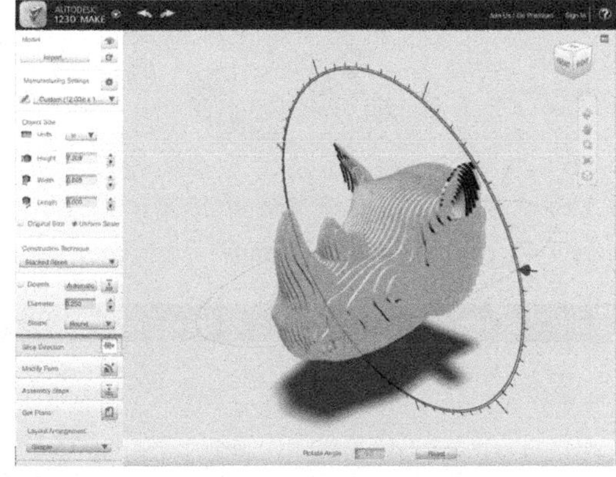

Figure 5-40 *...to adjust the direction of a slice*

Modify Form

Modify Form allows you alter your model in ways that render it easier to assemble after printing. The options include:

Hollow
: Makes your solid object hollow. This is a super useful feature that will save you material and expense. It can also make an

object more lightweight, or even floatable!

Thicken
: Makes a part thicker. If you find that areas of your part are a bit too thin for practical manufacturing, this tool will help you correct that.

Shrinkwrap
: Very similar to Thicken, Shrinkwrap adds a thicker skin to your part. For example, if your model includes lots of holes, Shrinkwrap will plug them up.

Figure 5-41 through Figure 5-43 is a ridiculous example that illustrates the point.

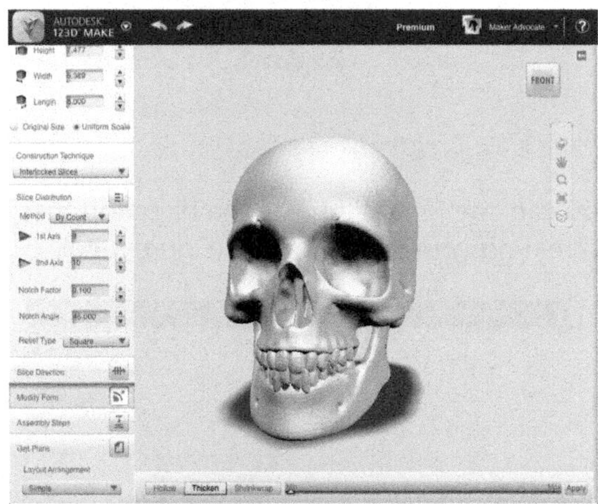

Figure 5-41 *Normal model of a skull*

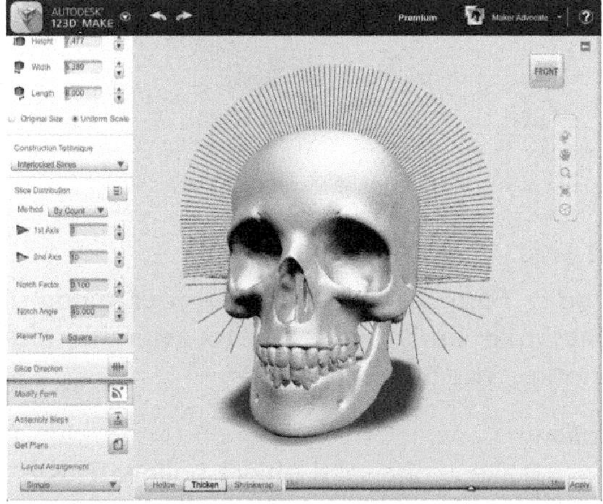

Figure 5-42 *The model preview using the thicken tool*

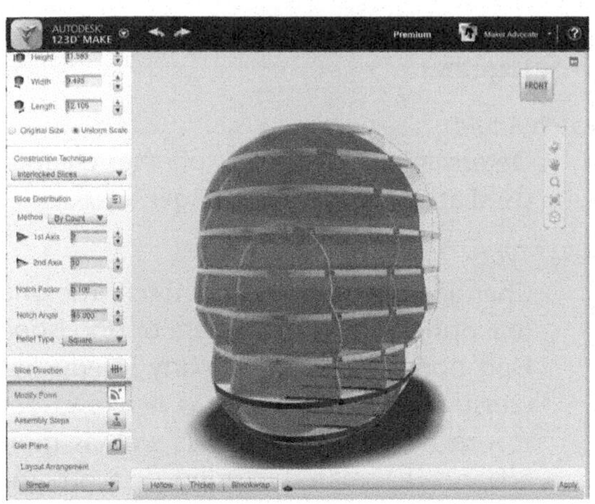

Figure 5-43 *The finished model after using the Shrinkwrap tool*

Assembly Steps

This tool will literally show you the best order of steps for assembling your model. When selected, click on the left- or right-pointing arrows on the bottom right side of your screen. This will walk you through your construction steps begining with the first two pieces that you should connect and in what order to add them. This is super helpful when choosing the interlocked construction technique, as it can get a bit tricky, even though everything is numbered. If you click on the left arrow, it will take you back a step, while the right arrow brings you forward.

Get Plans

You guessed it, this is how you actually leave the digital world and enter the physical. This takes your drawings that are shown on the right side of the screen and turns them into vectors that your cutting tool will follow in order to make each piece.

Options include EPS, DXF, or PDF. If choosing DXF, be sure to indicate what measurements you would like the resulting file to use, and be sure to clarify if you want the resulting file to use English or metric measurements (many cutters only have metric software). Then press the Export button to save your file in that format.

You will want to choose the Nested option under Layout Arrangement to save material and significantly shorten cutting time. If you are using a laser to cut out your objects, you can open up your DXF, EPS, or PDF file in your favorite vector design program like Corel or Adobe Illusrator and move your pieces to be almost touching. The reason for this is that lasers are precise enough that you can save a lot of material by moving your objects very close together. You would not want to do this on a CNC router unless you spaced them apart based on your router bit. If you are using an 1/8th-inch cutting bit and your objects are only 1/16th of an inch apart, then you will be cutting into your parts while cutting out others.

OK, now that you know how to get around 123D Make, you'll really appreciate some of the amazing things that people have created with it (see Figure 5-44 though Figure 5-47)!

You can find more information about these projects, including build instructions, by checking out Instructables.com.

Figure 5-44 *Solar hot-water heater (http://bit.ly/1NsE7j8)*

Figure 5-45 *Strata Bench by member ADintheStudio (http://bit.ly/1m1oLek)*

Figure 5-46 *Awesome Lamp by member PATHfab (http://bit.ly/1jYpRpG)*

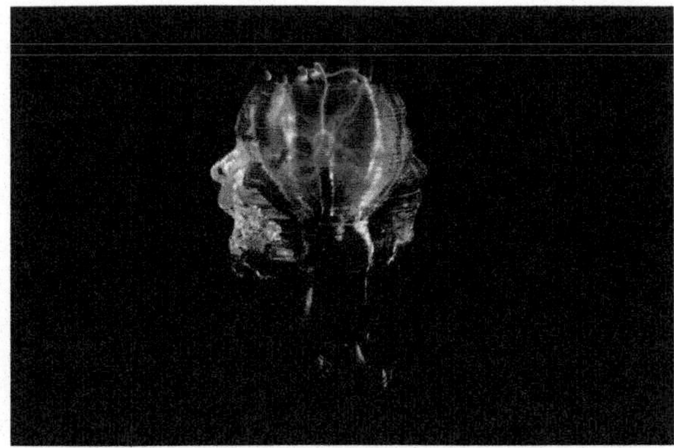

Figure 5-47 *Confusion Sculpture by member Anaisa-Franco (http://www.anaisafranco.com)*

6 123D Catch

123D Catch allows anyone to turn photographs of places, people, and things into high-definition 3D models—a process that used to be expensive and available to a relative few lucky professionals. These models can then be optimized for the Web, mobile, or 3D printing.

The Android and iOS versions of Catch allow you to capture images directly with your smartphone or tablet camera.

People have used Catch to capture objects of all shapes and sizes, from the Venus de Milo (Figure 6-1), to a sleeping cat (Figure 6-2), to standing stones in Scotland (Figure 6-3). It is a remarkable and powerful tool.

Figure 6-1 *A Catch 3D model of the Venus de Milo, by German Vargas*

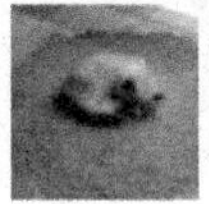

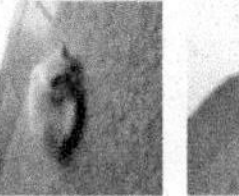

Figure 6-2 *A Catch 3D model of a sleeping cat, by Pierre-Olivier Mari*

Figure 6-3 *Bre Pettis, one of the inventors of the MakerBot, used Catch to create a 3D model of a standing stone in Scotland*

How Does 123D Catch Do It?

123D Catch works by reading the pixels that make up a digital photograph: the tiny dots of red, green, and blue.

As the photographer moves around an object to collect different views, Catch assigns each pixel in the image a unique code, such as R-40, B-578, G-564, and so on. The app stores the codes for those pixels that comprise the object, and deletes the codes that it applies to pixels in the background.

Catch works best with the 123D Catch mobile app, because the cameras built into mobile devices tend to have great "depth of field."

Depth of field is a photography term that refers to how much of what's in a camera's field of view is in focus, both in the foreground and deep into the background. Depth of field is important with Catch because you want your entire object to be in focus no matter how close or far apart it is from the camera lens.

However, there is a desktop version of Catch, called Memento, that allows you to stitch together photographs, or scans of photographs, taken with a nonnetworked camera, such as a digital SLR. The advantage of using these types of images is that digital SLR cameras can usually take higher-resolution photographs than mobile devices, which in turn means that you will get a cleaner 3D model requiring less clean-up work.

As this book goes to print, Memento is available in beta for the Windows platform, with a Mac version upcoming. Autodesk also offers a pro version of Catch called Re-Cap. You can find both of these programs through Autodesk.com

While catching 3D images of objects is neat, there are particularly innovative applications of the technology that involve fitting manufactured objects to organic objects, such as human bodies. This opens a new world of design potential for jewelry, fashion, and prosthetics: instead of playing a guessing game to make objects that fit a large number of people, 3D capture allows designers to create customized solutions for individual end users that fit more comfortably.

To see this in action, check out Dutch designer and maker Anouk Wipprecht's Faraday Dress. Wipprecht created a 3D scan of her own body with 123D Catch, and then laid out the 94 panels for her dress, as well as unique shoulder pieces, using two other apps: Autodesk Maya and 123D Make.

The end result was electrifying: a high-fashion Faraday cage (Figure 6-4) that allowed Wipprecht to stand between two Tesla coils and safely intersect with bolts of high-voltage electricity.

Figure 6-4 *Anouk Wipprecht in her Faraday Dress (photo by Kyle Cothern)*

Scanning with Catch: The Basics

Although 123D Catch is a free, easy-to-use app, there are several basic steps and techniques to keep in mind when scanning in an object that will help you achieve expert-level results.

Let's take a look at some tips and tricks:

Have a solid plan for your photo shoot
: Do a practice run to figure out how to get all the way around your object, while keeping it roughly in the same space in your camera's frame.

  To get the best result, we recommend doing a test run and marking the spots on the floor that you want to hit for the final capture.

Remember: The object does not move
: You do. The way photogrammetry works is that it reads pixel data. It matches up the pixels that read the same from shot to shot, and deletes the data that changes drastically.

  For instance, if you are using Catch to capture a 3D model of a person in a room, the software's algorithm would read small details—things like an eye, or a freckle on the person's cheek—and use them as anchor points. When you move about 15 degrees to take the next shot, the software will look for and read those same anchor points. It will then delete all the parts of the image that are in the background and out of focus.

Figure 6-5 *Image by Instructables user damonite*

Fill the frame
: Catch gets the best results when you fill the frame nearly edge to edge with the object, and then keep the object in that frame as you move around it (Figure 6-6).

Figure 6-6 *The dinosaur you wish to scan*

Move in a slight upward spiral or two loops

As you circle the object, move the camera in a slight upward swirl or spiral (Figure 6-7). Using Catch is like a dance: it's OK to take practice shots to find the rhythm and footwork that get the best results.

The app version will prompt you to take a bottom loop then a top loop. You will want to make sure that you are shooting at a slightly upward angle when doing the bottom loop and a downward angle when doing the top. The blue highlighted bar will tell you if you are lined up at the correct angle around the object.

Figure 6-7 *Image by Instructables user damonite*

Catch works best with ambient lighting and a busy background

No need to have an all-white photo booth and careful lighting to get great results. In fact, Catch works best with ambient and low-contrast lighting, such as that found on a cloudy day. It's also good to have a lot of detail in the background, so that Catch can tell which pixels to keep and which to throw away.

Because of the way the software works, a great environment in which to do a capture run would be a room with newspaper pages covering the walls, because they would create a varied background that the app could easily distinguish from the object you want it to focus on.

Dont forget the undercuts

Although it is now easier than ever to clean up scanned models when they have the occasional hole, the entire reason that you are scanning something is to get as much accurate data in the first place as possible. So if your object has some extreme undercuts, make sure you shoot the image from a vantage point that grabs those babies (Figure 6-8).

Figure 6-8 *Undercuts shown in white*

No hard shadows, no shiny edges

Because things like hard shadows and highlights move as you move, they typically don't work well for 3D scanning. If your object is naturally shiny, try applying a fine spray powder, such as the kind sold for foot odor. This will dull the object's finish so that you can capture a good scan, and it's easy to remove afterward (Figure 6-9).

Figure 6-9 *The results of shiny edges*

Let's Take Some Photos!

Now that you understand what Catch does, and how to use your camera to obtain the best results, let's put that knowledge into action and capture some data.

Step 1: Make a Practice Run

If your object is portable, set it up in a way that you can comfortably walk all the way around it. If not, be sure you that you can create a safe area to get around the object. Be sure to watch for people and vehicle traffic if you are shooting in a busy outdoor location (Figure 6-10).

Figure 6-10 *Clear a path before starting to shoot*

Holding your camera up and using the viewfinder, walk around the object once to make sure there are no literal stumbling blocks like walls or rocks that would stop you from a successful capture. Doing this before you make your scan helps reduce the need to make major adjustments to the object mid-shooting, which can complicate the creation of an accurate scan (Figure 6-11).

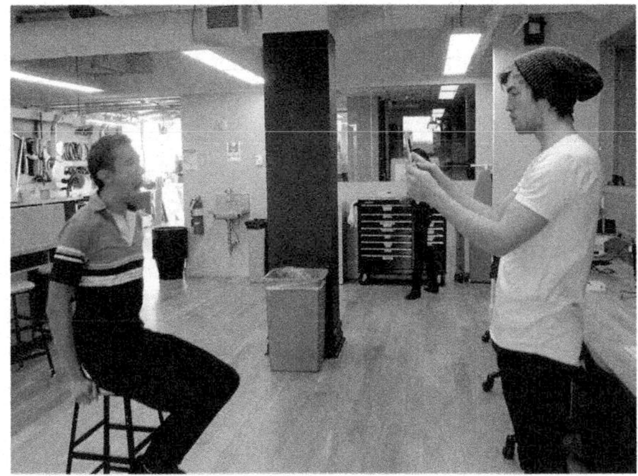

Figure 6-11 *Walk around your object*

Step 2: Check and Correct Lighting

Now check the lighting. It should be ambient and diffused, allowing you to see all the details of your object while creating no high-contrast shadows or highlights. The perfect scenario is an overcast day (Figure 6-12); this makes bright light without creating highlights. See if you can replicate this if you are in an indoor environment.

Figure 6-12 *Perfect lighting for a great capture*

Step 3: Capture the Object

If you are using the app version of Catch to create your photos, open that now. Swipe left to start a new capture (Figure 6-13).

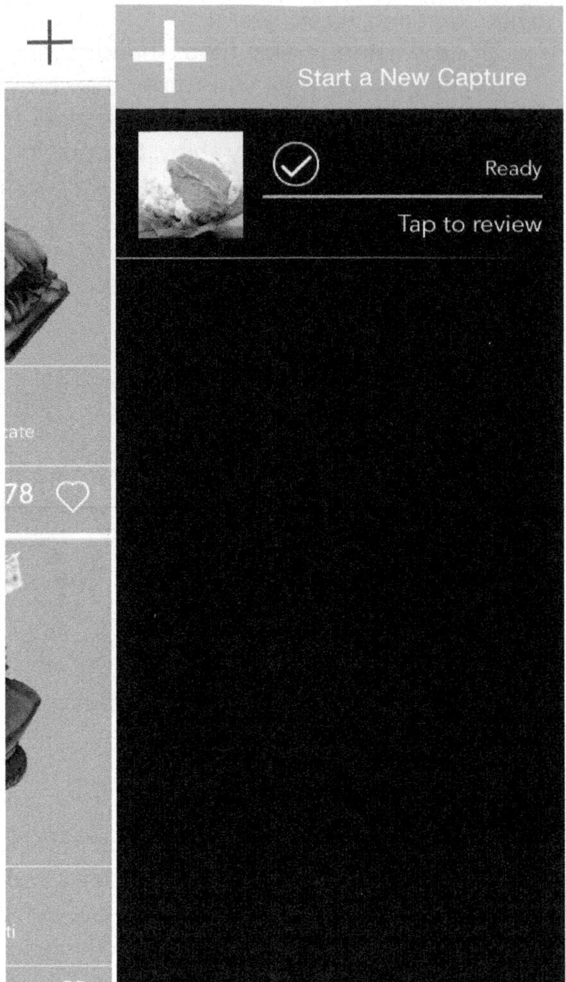

Figure 6-13 *The Start a New Capture button in the iPhone app*

If you're using a handheld camera: while looking through your camera's viewfinder, move closer to or farther from your object as needed, until it fills but does not overlap the frame.

Begin to move around the object slowly. As you slide to the next shot, keep the details from the previous shot still in the frame. You want to take a new photograph for every 15 degrees of movement (Figure 6-14).

Figure 6-14 *Seeing the path around the object*

Step 4: Prepare Your Shoot

At this point, you need to prepare your shot (don't shoot until the next step). Find your start point, usually the focal point of the object; this will be your home base. You are going to take about 20–40 photos, so it's OK to take your time. If you are using the mobile app, it will let you know how many photos remain to complete the scan of the object (Figure 6-15).

Figure 6-15 *The oscillating indicator level in the Mobile App*

You are going to take two laps around your object. Begin by taking pictures low on the object, with the camera facing slightly upward, on the side that has the most detail. For convenience, we'll refer to this as the "front" of the object (even if it is actually the back or side).

After you complete the first lap, start the second lap from the end point of the first lap. During this lap, try to get more above the object if possible, and tilt the camera slightly downward.

Step 5: Start Shooting

Start by taking a photo at the lower front of the object, which we'll call home base. (If you are only capturing a 180-degree object, your home base will be to either the far right or left.)

Take a picture and move about 15 degrees. Take your next shot. Repeat. You are doing 3D capture!

The 123D Catch mobile app prompts you by showing where your current view lines up with the blue marking at the bottom left of the screen (Figure 6-16). When you get back to home base (or if you are doing a 180-degree object, get to the far side), start doing your upper lap around the object.

When you get back to home base a second time, you are ready to process your image captures.

Figure 6-16 *The white cameras show you the appropriate path around an Object*

Step 6: Time to Process

If you are using the mobile app, click the blue checkbox on the upper right to finish capture. Then you can review your shots (Figure 6-17).

After review, click to start processing the capture.

Camera 23 photos Submit

Review your photos.
Tap to delete or retake.

Figure 6-17 *Capture of Catch Review on a mobile device*

Your capture might take as long as 15 minutes to process (because this all happens in the cloud). I highly recommend being on a WiFi network for this, rather than a cellular phone network.

If you are using a camera and a PC to create your 3D capture, open up Autodesk 123D and upload the photos that you've just made in order to process them. This process takes between 10–20 minutes depending on the size of the photos as well as server traffic; you'll be notified via email when it's done (Figure 6-18).

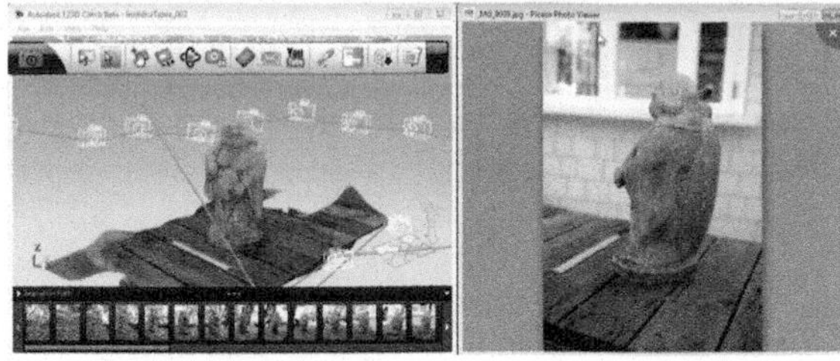

Figure 6-18 *Screen grab of what a finished Catch looks like on the PC version*

Step 7: Reviewing the Results

At this point, you should be able to see your capture in full 3D, either on the mobile device or on the desktop PC app.

If you are noticing a lot of distortion on an area, or a large empty spot on your object, try rescanning the object. One helpful technique is to put colored dot stickers or small pieces of painter's tape on trouble areas, so that you'll remember to capture them more effectively on the second try (Figure 6-19).

Figure 6-19 *Example of color marks on shoulders to get better results*

Now That I Have a Scan, How Do I Clean It Up?

The easiest way to clean up a scan is to use Meshmixer, another free app that you can find at 123Dapp.com (refer back to Chapter 3 for more information about Meshmixer).

Once you have downloaded and installed Meshmixer on your Windows or Mac computer, sign in to 123Dapp.com, download the model you want to work with from your account, and then import it into Meshmixer to preview, refine, and prepare it for 3D printing (Figure 6-20).

One approach is to use the Analyse → Inspector command in Meshmixer to find the gaps and holes in your object's mesh: its surface details. Use Meshmixer's tools to fill in the missing information and make your mesh watertight.

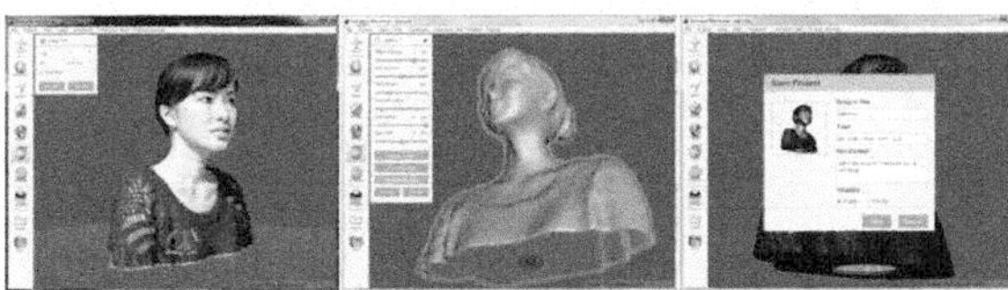

Figure 6-20 *Cleaning up a scan in Meshmixer*

How Do I 3D Print My Model?

Again, Meshmixer makes this pretty simple: you can customize and order a 3D print of your object from services such as i.materialise, Sculpteo, and Shapeways.

If you prefer, you can also print directly to your own MakerBot, Type A Machine, or other 3D printer.

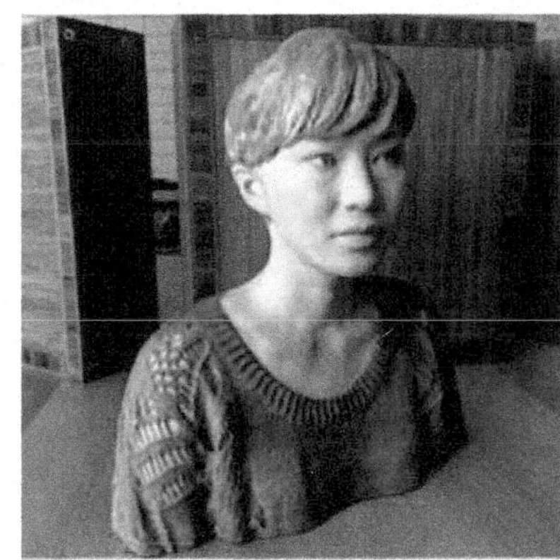

Figure 6-21 *Full color 3D-printed Catch*

What's Next?

Now that you have a primer on 123D Catch, you're ready to start scanning more objects! For more detailed tutorials and additional information, visit *http://www.123dapp.com/howto/catch*.

Index

Symbols

A

B

C

D

E

P

Q

R

S

T

U

Z

About the Authors

Emily Gertz is a journalist and the co-author of *Environmental Monitoring With Arduino* and *Atmospheric Monitoring With Arduino*, both published by Maker Media.

Jesse Harrington Au is the Chief Maker Advocate at Autodesk Inc. He is also a designer, maker, and educator. Au provides technical support both internally and externally to Makers, Fashion Designers, Product Startups, and Fine Artist.

Colophon

The cover image is from Jesse Harrington Au. The cover and body font is Myriad Pro; the heading font is Benton Sans; and the code font is Ubuntu Mono.